The Future of Brain Repair

The Future of Brain Repair

A Realist's Guide to Stem Cell Therapy

Jack Price

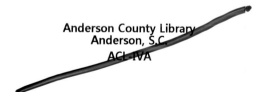

The MIT Press
Cambridge, Massachusetts
London, England

© 2020 Massachusetts Institute of Technology

All rights reserved. No part of this book may be reproduced in any form by any electronic or mechanical means (including photocopying, recording, or information storage and retrieval) without permission in writing from the publisher.

This book was set in Stone Serif and Stone Sans by Westchester Publishing Services. Printed and bound in the United States of America.

Library of Congress Cataloging-in-Publication Data

Names: Price, Jack (Neurobiologist), author.
Title: The future of brain repair : a realist's guide to stem cell therapy / Jack Price.
Description: Cambridge, Massachusetts : The MIT Press, [2020] | Includes bibliographical references and index.
Identifiers: LCCN 2019026654 | ISBN 9780262043755 (hardcover) | ISBN 9780262357890 (ebook)
Subjects: LCSH: Brain—Diseases—Treatment. | Stem cells—Therapeutic use. | Regenerative medicine.
Classification: LCC RC386.2 .P75 2020 | DDC 616.8/0427—dc23
LC record available at https://lccn.loc.gov/2019026654

10 9 8 7 6 5 4 3 2 1

Dedicated to the memory of my old friend
John "Stan" Sinclair,
who would have done great things had he been given the chance.

Once the development was ended, the founts of growth and regeneration of the axons and dendrites dried up irrevocably. In the adult centers, the nerve paths are something fixed, ended, and immutable. Everything may die, nothing may be regenerated. It is for the science of the future to change, if possible, this harsh decree.

Santiago Ramón y Cajal, 1928

Contents

Preface xi

1 **The Promise** 1
2 **Why Doesn't Brain Repair Work?** 19
3 **New Cells for Old Brains** 47
4 **Neural Stem Cells** 73
5 **The Cell Therapy Approach** 91
6 **Cell Replacement** 109
7 **Stem Cells Redefined** 129
8 **Feral Therapies** 147
9 **Pluripotency** 167
10 **Histogenesis** 191
11 **Mutation and Rejection** 211
12 **Prospects** 229

Acknowledgments 239
Notes 241
Index 265

Preface

In 1996, I received a visit from two academic scientists. At that time, I was Director of Molecular Neuroscience at SmithKline Beecham Pharmaceuticals, and the visitors wanted to talk about stem cell therapies for brain disorders. The visitors were from the Institute of Psychiatry in London, and they came to me because they hoped SB might invest in their stem cell technology.

They had chosen an unfortunate moment. I had just been part of a company working party investigating stem cell therapies. Upper management had heard of innovations in stem cells, and some competitors were moving into the area, albeit in a small way. Management wanted to know if there was any realistic prospect that stem cells might be used to treat major disorders, such as diabetes, heart disease, or stroke. They wanted to know if there might be an investment opportunity for our company in cellular therapy.

Our working party had talked to top scientists in the field in both Europe and the United States. We'd been impressed with the people and the progress, but we recommended that SB stay out of cell therapy. I had recently appeared on BBC Radio 4's *Science Now* to say that I couldn't see a breakthrough in stem cell therapy in Parkinson's disease—one of the main disorders under study—for at least five years. Much progress had been made, but I thought the challenges were still enormous. Furthermore, I had a strong suspicion that some researchers in the field, experts though they were, had underestimated the enormity of what they were trying to achieve.

I was trained as a developmental neurobiologist, and had been studying brain development for some years. Stem cell scientists were trying to make stem cells build new brain tissue, so they were trying to reproduce in the damaged brain what only normally happens during development. But brain development takes place in the very peculiar circumstances that occur only in the fetus, when the brain is much smaller and less complex. The idea that you could repeat the trick in the tangle of an aged brain seemed to me very unlikely—a bit like trying to walk across the modern City of London using the original 1875 Ordnance Survey map. Some of the routes would still be intact, but you would probably discover a few skyscrapers blocking your path.

So when I met my visitors—Jeffrey Grey and John Sinden—we talked back and forth for a while about what they were trying to achieve, until eventually they asked whether there was any chance that SB might be prepared to invest in their research. I said I doubted it. Why? asked Jeffrey. Because we don't think it will work, I replied. Oh, but we've already shown that it does, said Jeffrey, and he proceeded to describe a series of experiments they had completed. What the data seemed to show was precisely what I thought was impossible. Experimental animals with damaged brains had been injected with stem cells. As a consequence, new brain cells had been produced and—most important—the animals got better. The stem cells had apparently aided recovery from brain damage.

Psychologists have a term for the conflicted sensation I experienced at this point. "Cognitive dissonance" describes the disorientation you feel when your basic beliefs and assumptions are undermined by demonstrable fact. I probably looked as uncomfortable as I felt. Being an astute professor of psychology, Jeffrey recognized the situation and gave me his cheekiest smile. How about five million pounds, he asked?

Jeffrey and John didn't get their five million off SB, but the year after our meeting, together with another colleague, Helen Hodges, they formed a start-up company to commercialize their stem cell technology. The following year they asked me to join them, and I became a consultant to their new enterprise. Roughly a decade later, I helped them apply to the MHRA (Medicines and Healthcare products Regulatory Agency) for authorization to commence a clinical trial for a stem cell therapy in stroke. This became the "PISCES trial," the first clinical trial in Europe of a stem cell therapy for a neurodegenerative condition.

Interest in stem cell therapies for brain disease has grown substantially over recent years. I am constantly contacted by patients, or more usually by relatives of patients, anxious for news of a successful therapy. I have contributed to conferences and working parties in diverse settings—the London School of Economics, the Nuffield Council on Bioethics, the Academy of Medical Sciences—and been struck by the interest of diverse professional groups in stem cells, from sociologists to ethicists, anthropologists, and lawyers, all eager to understand how stem cell therapies might impact their domains. The general public is no less engaged. One of the most exhilarating discussions I have had on brain therapies was following a lecture to the "University of the Third Age" in Cambridge. Mostly retirees, with enormous depth and breadth of knowledge and experience, this audience had a particular interest in diseases of the aged, and they wanted to understand precisely the prospects for effective therapy. The stimulus provided by these diverse interactions represents the impetus for this book.

What follows, though not a memoir, is in a sense a progress report on my state of cognitive dissonance. Will there be successful stem cell therapies for the brain or won't there? If I was right about the likelihood of it working, how come Jeffrey's mice got better? If he was right, what on earth could the stem cells be doing—and might they do the same for human patients? The answer, for those who don't want to read to the end of the book, is that we were both right.

1 The Promise

Could we be about to experience a revolution in health care? If it happens, it won't be brought about by improvements in public health, better diagnoses, or more effective medicines. The advance will be the result of biotechnology. Biotechnology is not new, but the current pace of biotechnological progress is unique in human history. Some observers are of the opinion that the effect of this change will be revolutionary. Others believe it has potentially catastrophic potential for the future of mankind.

If we look at where biotechnology is likely to have its greatest impact, a broad range of possibilities emerges. We can foresee artificial organs, improved drug delivery, and personalized therapies. These advances will be innovative, though like most technology, they will be qualitative rather than dramatic. There are, however, two areas where we might imagine that there will be quantum change, where biotech could plausibly be transformative. One is the manipulation of the human genome; the other is the repair of the brain. The primary challenge in relation to the genome is how to safely alter our genetic constitution without compromising the genetic inheritance of our offspring. The challenge in relation to the brain is to rebuild brain tissue. In this book, I want to investigate the second of these, and examine the potential for a reconstructive therapy for the nervous system. Could stem cells deliver a true regenerative medicine of the brain? Is there truly a revolution in the offing, or a more modest advance? Indeed, will there be any substantial progress at all in treating major brain disorders?

The need for improvement in our treatment of brain disorders certainly exists. A comparison of the top three killer diseases in the Western world is revealing. The top two, cancer and heart disease, are slowly declining in incidence. This is largely the result of improved prevention strategies and therapies, plus the recognition of the importance of "lifestyle" choices. The United States and Europe, at least, are slowly weaning themselves off tobacco, and recognizing the important of fitness and diet. By contrast, in 2014, the third biggest killer—stroke—moved up to second place.

Ischemic stroke, the most common form of this disorder, is caused by a blockage of an artery that supplies blood to the brain. As a consequence of the blockage, an area of brain tissue is starved of oxygen and dies. The lost tissue might be the size of a pea or a plum but, either way, it has gone forever. If the stroke patient survives, the tissue will eventually be replaced by a fluid-filled cyst, water where brain should be.

Stroke is one cause of brain damage, but only one of several. Unlike stroke, where a discrete area of brain tissue is lost very quickly, cell loss in Alzheimer's is slow and diffuse, but equally permanent. Neurons—the information-transmitting cells of the brain—are outnumbered by support cells, the glia. But whereas lost glia can recover, neurons seldom do. In Parkinson's disease, cell loss is again diffuse, although the death of one specific population of neurons, called "dopaminergic cells," is strongly associated with Parkinson's cardinal feature, the loss of motor control.

These disorders are a growing concern in part because they particularly afflict the elderly and because they don't just kill, they disable. Three-quarters of strokes occur in the over-65s.[1] Roughly a quarter of strokes result in death within a year. Half of the survivors are left with disability. A third of these are disabled to such an extent that they require assistance with everyday activities. Stroke is in fact the single biggest cause of disability in the Western world.

The notion of a demographic time bomb is becoming increasingly familiar. People are living longer, which means the proportion of the population at risk of diseases of the aged is increasing. This results in more of us living with stroke disability, dementia, and the other consequences of brain tissue loss. A 2010 report suggested that the 36 million sufferers of Alzheimer's worldwide would increase to 115 million by the year 2050.[2] This anticipated increase is a direct consequence of an aging population.

In the face of this grim scenario, a case can be made for both optimism and pessimism. The case for pessimism is clear: conventional drug therapies are failing to respond to this challenge. There are both general and specific reasons for this shortcoming. Overall, the pharmaceutical industry is undeniably a highly inefficient business. On average, a novel drug takes a decade or more to develop. Worse, most new medicines fail. A clinical trial is the culmination of many years of preclinical research. Only the most promising new compounds are tried on human subjects. Even so, the failure rate in clinical trials is greater than 80 percent. The situation in relation to brain disorders is particularly disappointing. A 2014 review of novel Alzheimer's therapies concluded: "In the past 5 years, there have been 6 amyloidocentric programs that completed phase 3 clinical testing. None met their primary outcome measures."[3] Note that these are failures at the advanced phase 3 stage of clinical trials. After years of effort and millions of dollars of investment, these new therapies still didn't work. Further, the technical term "amyloidocentric" is revealing. Alzheimer's has been intensively researched in recent years, and volumes of data have been amassed to support theories into the cause of the disease. One favorite among these holds that a substance called "amyloid" accumulates around cells in the brain, poisoning and killing them. So if this buildup of amyloid could be reversed, the theory goes, then the progress of the disease could be halted. This amyloid pathology has been modeled in animals, meaning that the putative human disease mechanism has been mimicked in a living mouse. Significantly, all of the failed compounds would have been tested—and worked—in the mouse model. Indeed, success in treating experimental animals would have been a prerequisite for the new medicines to reach clinical trials. So, not only have the individual drugs themselves failed, the whole amyloid theory, and the animal models to test it, have proved unreliable, meaning that the whole field now has to look elsewhere.

Unfortunately, this predicament is not limited to Alzheimer's research. Stroke, as we have seen, is caused by the blockage of a major blood vessel the brain. That blockage is most commonly found in a vessel called the "middle cerebral artery." Again, this pathology can be mimicked in rodents. If the middle cerebral artery of a rodent is blocked experimentally, the animal suffers a stroke broadly similar to that seen in human patients. In the last twenty years, multiple novel compounds have been tested to see whether they could protect rats from brain damage following blockage of this artery. Several compounds succeeded in the experimental model, only to fail spectacularly in clinical trial. Malcolm Macleod and his colleagues in Edinburgh have studied the reasons for this failure. They observed that: "Despite billions of dollars spent, not one pharmacological agent, improving outcome in stroke patients by acting on brain cells, has been approved by the [US] Food and Drug Administration (FDA)."[4]

So, if you are of a pessimistic bent, you might conclude that we are in a quandary. The impact of brain damage on our population is substantial and growing, yet conventional drug discovery appears ill equipped to meet the challenge. In truth, the true picture for drug discovery is less bleak than this narrative suggests. Researchers are very conscious of these failures, and are working hard to rectify the causes. Also, novel approaches to early detection and alternative drug discovery strategies are showing considerable promise. But let's leave this story here for the moment and ask what possible grounds there might be for optimism. Which brings us back to biotechnology.

Humans have long manipulated biology to their advantage. Ever since we learned to herd animals and plant seed, we have used biology to improve our diet, our productive capacity, and our physical well-being. Since the turn of the twentieth century, however, the pace of change has increased substantially, and nowhere is this more clearly seen than in health care. The discoveries of Louis Pasteur and Joseph Lister gave rise to infection control, with profound impacts on public health and the management of infectious diseases. Before these advances, surgery was so risky that "only conditions that brought patients near death warranted the risk of surgical intervention."[5] Insights into physiology and biochemistry led to the discovery, purification, and synthesis of hormones and vitamins. In 1917, 60 percent of diabetic children died within 18 months of diagnosis. Now, thanks to the discovery and manufacture of insulin, such children in Western countries can expect to live out a normal life span. Before the Second World War, forty of every thousand women died in childbirth in the Western world.[6] By 1970, that number had fallen to fewer than one in a thousand. Antibiotics and immunization brought such rapid advances in public health that, within a generation, some of the great killer diseases—polio, smallpox—all but disappeared from human populations. Indeed, nothing we can envisage for the future is likely to rival the achievements of biomedicine up to 1960.

During the final quarter of the century, biomedical advances became more complex and more technologically sophisticated. New imaging techniques were discovered to aid diagnosis. Keyhole surgery reduced the trauma of operations. Cochlear implants gave hearing to the deaf. In most cases, these advances were almost invisible outside of the biomedical community, a good example being the advances in diagnosis and disease monitoring through a complex array of laboratory tests and assays now available to clinicians. But late twentieth-century medical technology has also permitted an entirely new development: personalized medicine of breathtaking complexity. Consider the headline biomedical breakthrough of 1978—in vitro fertilization (IVF). Eggs are taken from a woman's ovaries. Sperm is harvested from her partner and either fertilizes the egg in a dish, or is injected into the egg. The fertilized egg is then cultured in an artificial medium for 5 or 6 days, then finally the new embryo is introduced into the mother's uterus, where it implants and develops to term. To aid implantation and normal development, the mother is injected with hormones to mimic the early stage of pregnancy. Remarkable to think that just fifty years before this innovation, doctors struggled even to keep women alive during childbirth. Now a prospective mother can expect to have this panoply of biotechnology and knowhow at her disposal, not to save her life, nor even to aid her unborn baby, but merely to overcome her infertility, a distressing but hardly life-threatening condition. IVF technology is far from perfect, of course: the failure rate is high, and women often endure multiple rounds of hormone treatment. Nonetheless, to the women of the 1930s, IVF would have seemed like magic, and to some extent it still does.

In the last decade, the trend has continued for biotech solutions to become more intensely technological, more personalized, more fantastic in ambition. A whole new lexicon has emerged to describe these innovative technologies. We now speak of "stem cells" and "regenerative medicine" with familiarity if not understanding. Newspapers report on progress toward "designer babies" and on new genes "for" another human disorder. Educationalists worry about "cognitive enhancers," and we all wonder what to think about "lifestyle drugs."

Consider this report, published in the *Lancet* in 2008: a young woman had her bronchus—the airway connecting the windpipe to the lungs—surgically removed, then reconstructed using a donor's windpipe together with the patient's own stem cells.[7] Think of it: a diseased bronchus was excised from a living patient and the windpipe taken from a deceased donor. The donor's cells were then removed, and fresh cells (of two types) were cultured from the patient's own bronchus and bone marrow in order to generate the cells lining the airway and the external structural cells supporting the windpipe. Finally, the bioengineered structure was surgically implanted into the patient (a young mother in her thirties), who was restored to health, at least for the four months of the study. Hard to imagine that the doctors, scientists, and technicians who achieve successes such as this couldn't do anything they set their minds to. Yet this study, which gave rise to what the press would call "the Macchiarini case," turned out to be profoundly flawed. Both the study's patient and a further two patients died shortly after undergoing the procedure, and the surgeon involved was subsequently investigated for malpractice and dismissed by his institution.[8] A reminder, if any were needed, that hubris and overconfidence can flourish perfectly well in the biomedical community.

Certainly this feels like a biomedical revolution, but also a little scary. "Drugs Treat Symptoms. Stem Cells Can Cure You. One Day Soon, They May Even Stop Us Ageing," reads the subhead to a 2009 *Guardian* article. The authorization to begin a clinical trial of a stem cell therapeutic was hailed as "an extraordinarily exciting event [which] marks the dawn of a new era in medical therapeutics" by Thomas B. Okarma, the CEO of Geron Corporation, the biotech company developing the therapeutic.[9] Others warn that the latest development in genetic engineering "brings designer babies one step closer" and "will realize our greatest fears" of dehumanization. The twin themes of threat and promise are rarely far apart, and for those inclined to disbelief there is the salient feature that both proponents and opponents of the biotech revolution seem to agree on its potential. Nancy King and Jacob Perrin agree that stem cell therapies "hold great promise," but warn that "moving forward with the right blend of creativity and caution is essential, in the interest of both science and patients."[10] It seems the issue is not whether biotechnology will revolutionize our lives, but whether we should fear or welcome this revolution.

It is remarkable how rarely either advocates or opponents of the new biotechnologies describe in any detail the path from where we are now to the remarkable future they predict. The technologies themselves are most often referred to generically: "cloning," "genomics," "stem cells." We assume that the assertions these commentators make are backed up by a substantial technical literature, and that we too would be able to follow the technology if only we could "speak science." But scientific publications deal only with what has been achieved, not with what might be. The future directions are in the minds of the observers, not in the technical accounts. So, we seriously need to ask: How do we get from here to there? If we are going to regenerate our failing bodies, how is that going to come about?

The metaphor of stepping-stones or a ladder is common. Francis Fukuyama refers to a "ladder of complexity."[11] Biotechnologists will solve the easy problems first, then simply work their way up the ladder of complexity to address the more difficult issues. Biotechnology is a sort of escalator, it seems, advancing us toward ever more ambitious technologies. "It is conceivable that over time we will gain enough information to control the behavior of every cell in our bodies," observed biotechnologist William Haseltine in 2002. "Once we have achieved such mastery, we will be able to heal any disease."[12] This capacity for immortality will simply follow from the trajectory we are currently following. Again, note that Haseltine is broadly in favor of, whereas Fukuyama is broadly against, but both use similar metaphors, both see a brave new world emerging stepwise from the inherent progress of biotechnology. An interesting question is where this particular image comes from and why commentators with otherwise different standpoints adopt a similar perspective. Some scientists themselves nowadays espouse this view, the result perhaps of an environment in which science increasingly has to justify itself in terms of deliverables.

Of course, scientific progress only looks smooth if we look backward. As the Russian revolutionary thinker Alexander Herzen remarked, history has no libretto. In retrospect, we can see genetics progressing from Mendel's peas, through Watson and Crick's double helix, and forward into modern genomics. Similarly, we see stem cell technology emerging from progress in embryology plus the biomedical imperatives of transfusion scientists and radiobiologists. But the unhindered inevitability of progress is illusory. Scientists themselves—who tend to look predominantly forward rather than backward—are more acutely aware of the unpredictability of progress, of the cul-de-sacs, the failed endeavors, the attractive hypotheses that prove false. They see the no-go areas, where even brave investigators fear to tread. They have experienced the false dawns. If future science is at all like past science, many of the avenues that now seem open and sunlit will in fact lead nowhere: the path of stepping-stones will run out halfway across the water; the ladder reach only partway up the wall.

So we have an interesting juxtaposition. For more than a century, our lives have become steadily healthier and longer as a direct consequence of biomedical progress, progress that is becoming so sophisticated and complex that its recent successes seem barely believable. Some see this trajectory continuing over the horizon into the future. We just have to keep climbing that ladder. We cured the old killer diseases; now we have to address the new killers.

Except, some uncomfortable facts emerge. Despite the breakthroughs that brought us past successes, few new therapies for brain disorders have appeared. This has not been for lack of effort. In 1990, George H. W. Bush announced his "Decade of the Brain," in which "a new era of discovery is dawning in brain research." Similar quotes accompanied Barack Obama's "BRAIN initiative" in his 2003 State of the Union address. Neuroscientists haven't been sitting on their hands: they can point to great successes in neuroimaging, molecular genetics, and novel neurotechnologies. Nonetheless, the broad objective was largely similar in both these presidential initiatives, separated though they were by thirteen years. This feels less like ascending the ladder of success and more like a possible roadblock.

Cause for Optimism?

Stem cells might provide a way forward. The rationale for believing this is simple, at least in principle. Stem cells have the potential to generate new cells of different types, including new brain cells. We've seen with the windpipe-replacement story that stem cells can help build new tissue, and as we progress through this book, other such examples will arise. So could it work for brain tissue? If stem cells were injected into the brains of individuals who had had lost brain cells as a consequence of disease, then perhaps the stem cells could replace the lost cells. And if these new brain cells could be correctly wired into the brain, then in theory at least they might be able to restore function.

This simple idea quickly gets complicated when one considers the diversity of stem cell types, the variety of types of damage that disease can cause, and the logistical constraints around cell supply and delivery. Nevertheless, there is a conceptually simple experiment that has actually been successfully performed, in one variant or another, many dozens of times in animal model studies of stroke, Huntington's disease, traumatic brain injury, and other neurodegenerative conditions. A hole is drilled in the skull, and a slurry of stem cells is injected into the brain. Remarkably, when stem cells are transplanted into damaged brain tissue in this way, they do indeed bring about some functional recovery. Moreover, it works consistently and robustly—at least in experimental animals. We've had cause to doubt the veracity of animal models once already in this chapter, and this theme will reoccur. Nonetheless, these studies are pivotal in justifying the whole stem cell approach to brain repair.

For the sake of simplicity, let's just consider one variant of our experiment, that concerning stroke. One of the primary blood vessels of the brain is blocked in an experimental animal thereby closely mimicking stroke pathology in humans. The animal is usually a rat or mouse, but larger species like primates have also been studied. This blockage prevents nutrients and oxygen from reaching part of the brain, and just as in humans, that tissue can no longer maintain itself and quickly dies. The animal loses sensation and mobility in one side of its body, and this produces symptoms of functional loss that would be familiar to anyone who has seen a friend or relative after a stroke. Again as in human stroke patients, the disability in the animal is pretty stable. There might be some immediate improvement as the animal accommodates to the injury, but typically the disability remains for the four or five months that such an experiment usually extends.

The experimenter then injects a population of stem cells into the brain of the afflicted animal, usually close to the site of the stroke. Some researchers have injected the cells immediately following the stroke; others have waited to allow the disability to stabilize, thereby mimicking the chronic disability of human stroke patients. The experimenter will monitor evidence of change in the injected animals compared to control animals, a group that have also suffered an induced stroke but have received no stem cells.

Some researchers have looked primarily at whether the injected stem cells have given rise to new neurons in the brains of the stroked animals. Results here have been both quite variable and quite contentious. Some researchers think they see new neurons in the engrafted brains, whereas others have questioned whether the cells they observed were really new brain cells, and whether there were enough of them to make any difference.

More consistent data have emerged, however, when researchers have looked for functional improvement in the injected animals. These investigators, often clinician-scientists or scientists with training in psychology, have argued that it is not what happens to the injected cells that matters, but what happens to the animal itself. Is the animal better able to move? Does it have more control over its limbs? After all, this is what they are looking to achieve eventually in human patients. Stroke sufferers want to recover the use of an affected hand. They want to feel less fatigued. And if at all possible, they woud like to be able to get out of their wheelchairs.[13]

When the researchers look for this kind of improvement in the engrafted animals, consistent results do emerge. Over a large range of studies, engrafted animals show improvement. This can be seen in different ways. For example, the experimenter can put the engrafted animal on a narrow beam and measure how long it takes to cross to the other end, how many foot faults the animal makes, and how often it falls off the beam. A particularly useful measure is the "sticky tape" test. The experimenter puts a piece of ordinary parcel tape around each of the animal's forepaws. A rat would normally rip this tape off pretty quickly, but a stroked animal has problems sensing the tape on its afflicted side, and because of the loss of mobility, has problems removing it. Hence the stroked animal rips the tape more slowly off the affected paw than off the unaffected one. That being the case, the experimenter asks, can the injection of stem cells remedy this asymmetry?

The consistent finding from such experiments is that the animals that have been injected with stem cells are faster across the beam, have fewer foot faults, and fall off the beam less often than animals that have suffered the stroke but received no stem cells. In the "sticky tape" test, the animals that have received the stem cells now pull the tape off the affected paw as quickly as from the unaffected one, just as they did before the stroke.

There are two points to make about these observations. First, this is a robust finding. Scientists generally don't like to trust any result, particularly an important result, unless it has been replicated by several researchers working in different laboratories. This result has proven robust in exactly this fashion. Second, this result is decidedly nontrivial. There are not many agents that when injected into an animal with brain damage can bring about this type of functional recovery.

Such a result not only raises the question of whether stem cell engraftment could bring about a similar recovery in human patients, but also makes a fundamental point— that repair of damaged brain tissue *is* at least possible. Of course, these are rats not humans; they are otherwise young and healthy, unlike the aged humans who typically suffer strokes; and this is controlled experimental brain damage, rather than the complex multisystem dysfunction that typically accompanies human stroke. Nonetheless, this result shows that functional improvement following brain damage can be achieved in a mammal, albeit one separated from humans by 75 million years of evolution. For me, the important take-home message is that if we can work out what is happening in these recovering animals, we might be able to bring about a similar recovery in human patients. Stem cell therapy itself may or may not prove viable, but if we can work out what the stem cells are doing in the brains of these animals, we may finally have an effective therapy for stroke and similar brain disorders.

The more immediate hope, of course, is that this result may be directly reproduced in human patients. If it works in rats, why wouldn't it work in people? This is the question I want this book to address.

Unsurprisingly, we'll discover that this simple question has a complicated answer. There are features of the human brain that make repair difficult, the first being quite simply its complexity— and not just its obvious structural and functional complexity, but also a developmental complexity. The intricate and convoluted processes that the brain uses to build itself in the first place cannot easily be replicated therapeutically. As impressive as the technical wizardry of IVF is, inducing human brain cells to express themselves in artificial circumstances is infinitely more demanding.

A second feature that makes brain repair a challenge is that the brain is actually pretty poor at repairing itself. Some tissues—liver, blood, and skin —repair themselves quite well, which means therapy need be no more ambitious than to help them do better. But the mammalian brain didn't evolve like that, though other animals we condescendingly call "lower vertebrates"—amphibians and fishes—repair their nervous systems much better than we do.

These considerations lead to the conclusion that the brain is special when it comes to damage repair. It isn't that neuroscientists haven't got far enough up the ladder yet to address this set of problems. Neither is it a shortage of good ideas of how repair might be brought about. Rather, the challenges to be overcome in rebuilding brain tissue are immense. Rebuilding adult brain tissue might even be literally impossible.

We might want to consider whether repair is even a good idea. If your computer wouldn't boot up, it would be inadvisable to go into the hard drive with a soldering iron. Similarly, some commentators think we should consider that interference with the delicate circuits of the human brain might do more harm than good. There are instances where the engraftment of new cells into the nervous systems of animals has caused epilepsy and death. We mess with the brain at our peril.

Ignoring that advice, we'll consider what might be achieved by regenerative medicine. Specifically, could we reproduce the rat experiment in people? We'll discover that this is not actually a new idea. It has been under investigation at least since the 1970s. We'll discuss why the idea might work, why success so far is limited, and why it is such a controversial approach even among brain scientists themselves.

Looking ahead, we'll discover a number of key features that emerge from this account as it unfolds. First, we'll discover that there has indeed been substantial progress in the last decade in devising regenerative therapies for brain disorders. There are now cell therapies in the pipeline that might be expected to deliver genuine efficacy in treating some neurodegenerative conditions. We'll also discover, however, that there is an enormous tendency to exaggerate the significance of new developments. It is rarely the case that current technology has achieved as much as its advocates celebrate or its opponents fear. The reasons for the hyperbole are probably the same as in all human endeavor: vested interest and ignorance. It suits individuals and organizations alike to exaggerate achievements for financial or reputational advantage. The temptation is particularly great in areas, like biotech, that most people understand poorly. Commercial enterprises are always assumed to have given in to this temptation, while academics were supposed to stand above the fray. Still, all universities now have publicity officers and technology transfer departments that measure success not in terms of veracity, but in newspaper headlines and patents secured. Ignorance is revealed both by the journalists who, intentionally or otherwise, amplify the extravagant claims and by the scientists themselves, who wander outside their comfort zone. One consequence of this hype is the growth of "stem cell tourism," where patients at the end of their tether travel overseas in pursuit of radical, untested therapies. This is another issue to which we'll return.

If stem cells provide a case for optimism that brain repair might be achievable, it is a conditional optimism. Primitive brain repair might be possible if we can assemble a genuine therapy from the bits and pieces of puzzle that currently lie on the table. A breakthrough, or rather a cluster of breakthroughs, is required for a true regenerative medicine of the brain to emerge. As ever in science, it is difficult to see from where this might emerge, but, at this moment, one development in the pipeline stands out as a potential game changer.

In 2006, Shinya Yamanaka at Kyoto University discovered a new way of making stem cells. Not any old stem cells, these were "pluripotent" stem cells.[14] A proper discussion of pluripotency will have to wait until chapter 9. Suffice to note here that this discovery has transformed stem cell biology. Whereas, previously, human pluripotent stem cells could only be generated from human embryos, with all the ethical and logistical constraints that such an approach entailed, now such cells can be made from skin, hair, blood, or even urine. This discovery, for which Yamanaka justifiably shared the 2012 Nobel Prize in Medicine, allows us finally to imagine that we might build human tissue from scratch. In the few years since Yamanaka's discovery, this has already been achieved, and whole brains—albeit small, misshapen, and underdeveloped—have now been generated in a tissue culture dish from these pluripotent stem cells.

One conclusion emerges already: the "ladder" metaphor will not do. When we look forward from where we are, we can't anticipate a stepwise progression to success. Rather, we see substantial obstacles that may or may not be overcome. Think of salmon swimming upstream. A fish can go a long way, swimming hard and making progress, without actually having to overcome any serious obstacles. Then, round the next bend lies a waterfall—with a bear sat on a rock. A lot of salmon are not going to make it up that waterfall. Just like science in retrospect, the path to the top might be clear once you get there and look back, but it's not clear from where we are now who will make it through.

2 Why Doesn't Brain Repair Work?

Why is brain damage such a problem? Lots of tissues get damaged during a lifetime, yet they seem to repair themselves without too much trouble. Cancer patients are often treated with cytotoxic drugs to reduce their tumors. Unfortunately, such drugs also kill the lining their gut. Once the treatment ceases, however, the gut recovers and the cell lining is restored. Similarly, when patients with liver cancer have their tumors surgically resected, a sizable piece of liver is cut away with the malignant tissue. Nevertheless, with time the liver regrows. Another example familiar to us all is blood. If you cut your finger, you lose blood, but of course it gets replaced. So confident are some people in their ability to replace lost blood that they regularly donate a pint for the benefit of others. So why is the loss of brain tissue so different? Why can't we *just regrow* brain cells?

First, we need to recognize that there are many ways to damage the brain and that the consequences of each are different. Mild brain damage, such as a concussion, might involve bruised brain tissue and ruptured blood vessels. While multiple concussions might lead to long-term disability—as many boxers and American football players know to their dismay—a mild concussion is usually followed by complete functional recovery, indicating that the brain does have at least some capacity for repair. More extensive damage to the brain can vary enormously: a gunshot wound to the head is quite different from a stroke, as is spinal cord injury from Parkinson's disease. Nonetheless, there are common themes that are central to the brain's failure to recover naturally. Let's take ischemic stroke as an example.

The Challenge of Stroke

In 2014, the British National Health Service (NHS) initiated a campaign to increase awareness of the signs of stroke (figure 2.1). The campaign was titled "Act F.A.S.T.," "F.A.S.T." being an acronym for "face-arms-speech-time," highlighting the most recognizable symptoms of an acute stroke: the sagging face, the fallen arms, and the slurred speech. Like the acronym itself, the fourth element of "F.A.S.T."—time—accentuates the urgent need to treat acute stroke

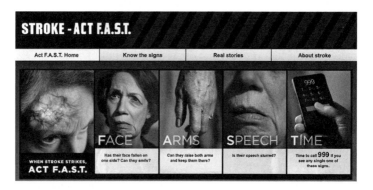

Figure 2.1
"Act F.A.S.T." Stroke publicity campaign poster, as displayed on the UK National Health Service website: http://www.nhs.uk/. Copyright NHS England. Used under terms of Open Government Licence. http://www.nationalarchives.gov.uk/doc/open-government-licence/version/3/.

as soon as possible. Current therapies for stroke are only effective if administered within the first few hours of the onset of attack. Unless immediate medical treatment is provided, the stroke will follow an unremitting course, frequently with fatal consequences.

The TV ads publicizing the campaign were very evocative. You see a middle-aged woman deteriorating before your eyes as a fire burns inside her skull, a potent image of the destruction being wrought in her brain. Fire is an appropriate metaphor. A forest fire can start with a spark yet spread rapidly. Elements that moments before were part of a balanced ecosystem are suddenly fueling a blaze. The spark, in the case of stroke, is the blockage of a major artery to the brain. Typically, the artery becomes obstructed by a small blood clot (figure 2.2). Immediately, all the brain tissue downstream of this blockage is at risk. Brain cells consume a high level of glucose and oxygen, both of which are transported around the body by the blood. In fact, even though the brain represents only about 2 percent of the body's weight, it receives about 15 percent of the blood flow from the heart and consumes 25 percent of the body's oxygen supply. Brain tissue is highly reliant on glucose metabolism, and the loss of its blood supply has immediate consequences.

In normal brain function, each neuron acts as a small capacitor and can discharge energy. There is a difference in electrical potential (voltage) between the inside and outside of each cell. If triggered to discharge, this voltage fires signals—known as "action potentials"—that travel along the nerve fibers. These impulses are the brain's information transfer system at work, and they underpin all brain function. Maintaining this voltage difference across the cell membrane requires pumps to continuously drive charge in and out of the cell. These pumps are high-maintenance and use up a considerable amount of energy.

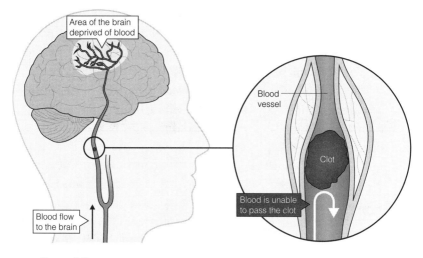

Figure 2.2
Ischemic stroke. A major artery supplying the brain becomes blocked, typically by a blood clot, depriving part of the brain of its blood supply.

Once the blood supply fails as in a stroke, the pumps stall, the ionic gradients are lost, and nerve function is compromised. Worse, the cellular biochemistry that drove brain function now turns on itself and starts to drive destructive processes, a condition called "excitotoxicity." As the pumps seize up, calcium ions flood into the cells and provoke the cells to release glutamate, a neurotransmitter whose normal job is to stimulate other neurons in the network. But now, in an uncontrolled rush, excitotoxic glutamate overstimulates neighboring neurons, spreading the flame. Astrocytes, the support cells of the nervous system, would normally take up this glutamate to protect the neurons, but, now overwhelmed, they only contribute to the damage. The inward rush of calcium also badly impacts neurons themselves: water follows the calcium, making the cells swell and distend. The calcium also stimulates a whole series of damaging enzymes, which start to break down the cellular infrastructure. Toxic chemicals called "free radicals" are produced, which damage membranes and cellular integrity, adding to the mayhem. The mitochondria, the cells' energy factories, are among the worst-affected cellular components, and their stress causes a bioenergetics failure. Within an hour of the artery blockage, neurons are seriously compromised and struggling to recover.

This gathering inferno attracts the firefighters. The damaged brain cells start to release "help" signals called cytokines. These messengers warn the rest of the body of the crisis, and summon blood-borne immune cells to the scene. First to arrive are the neutrophils. They force their way through the blood vessel walls into the brain tissue, engulfing debris and releasing further cytokines. Over the next hours and days other cellular components of the immune system will also engage: macrophages, monocytes, and lymphocytes. Concurrently, the brain's own macrophages, called microglia, become activated adding to the complex soup of factors and cells.

How the carnage is resolved depends on multiple factors. If the blood supply can be restored quickly, the neurons may yet recover. The aim of some current stroke therapies is to assist this recovery. Drugs called "thrombolytic agents" are used to dissolve the original clot, thereby restoring circulation; other approaches seek to physically displace the clot. If administered soon enough, these treatments can be effective. Without such intervention, however, the outcome depends on the size of the ischemic area, and and whether any blood at all is reaching the vulnerable brain. If a major vessel is blocked, affecting a large area of the brain, the recovery prospects for tissue at the core of this region are not good. Over the ensuing hours and days, the threatened neurons will succumb to the ischemia and simply die. The invading macrophages will be reduced to cleaning up the debris, and the integrity of the tissue will be lost.

Further from the core of the stroke, in an area referred to as the "penumbra," the outcome of the firefight is less certain. Multiple blood vessels serve the brain, and even though one is blocked, enough circulation might reach this penumbra to maintain cell survival. The extent of this collateral circulation will be an important factor in determining the degree of tissue loss. Individual neurons will battle to reestablish their integrity. If enough blood gets through, if inflammation can be contained, if the protective functions of the microglia can be sufficiently engaged, then neurons in the penumbra will survive. But where a balance cannot be maintained, cells will die.

It is a strange fact that all our cells have an intrinsic suicide program, a process called apoptosis.[1] They have mechanisms to monitor their own integrity, and if the cells fail to come up to scratch, these mechanisms will trigger cellular suicide. This type of cell death involves the activation of destructive enzymes, and the final outcome is that the cells eat up their own proteins and DNA, until finally they collapse, are engulfed by the invading macrophages, and disappear. The teleological explanation for how such a self-destructive program could have evolved is that it is better for tissue as a whole if damaged cells eat themselves up, rather than rupturing and spewing their contents into the surrounding tissue. Apoptosis is thus less inflammatory than necrosis, the uncontrolled cell death that predominates in the core of the stroke.

By this stage—hours or days after the initial blockage—about one in eight stroke sufferers is dead. If the stroke involved a critical brain region, or if it was too large, then the loss of brain function is simply incompatible with life. For the survivors, the tissue in the core of the ischemia has been lost, but in much of the penumbra the battle to save tissue has been successful. The defunct cells have been cleared away following the stroke, and the dead area sealed off.

A normal healthy brain is surrounded by a barrier called the "glial limiting membrane," which extends around the outside of the brain beneath the skull. This barrier is formed by astrocytes, star-shaped support cells of the brain. Astrocytes have blunt-ended extensions, which spread to the surface of the brain, where they come together and interweave to form a shield, like medieval infantry facing a volley of arrows. Following the loss of tissue in the stroke core, a similar barrier forms to seal off the lost area. So, the lost tissue becomes a fluid filled space, within the brain, but outside this limiting membrane.

The Outcome of Stroke

Where does this leave the brain, weeks and months later? Well, clearly, a piece of brain is now missing and all the functions it would have carried out have been lost. Brain plasticity might lead to the recovery of some of those functions, but the brain tissue itself will not recover. Nevertheless, this is not the full extent of the loss. To understand the full impact, we need to consider how the brain is built, and what makes it unique among the tissues of the body

From a cell biological perspective, the brain is unique in two regards. First, brain tissue is composed of a vast number of cell types. Counting cell types is, of course, an arbitrary business as the observer decides what constitutes a "type." But whereas most tissues are made up of a small number of cell types—let's say fewer than ten—the brain has more cell types than we can reliably count. In fact, as we'll discuss later, a case to be made that every pyramidal neuron in the cerebral cortex is unique.

Second, the number, precision, and complexity of interactions between cells are considerably greater in the brain that in other tissues. Cells in the body need to communicate, like members of any social group, and just like human interactions, some are near and some are far: some are fast and some are slow. You need to interact with the other people in your own home and your immediate neighbors. But you also need to know what's happening farther afield, at the grocery store, on the freeway, in D.C. You want to know what's happening now, but also what's planned for next week, for next year. So, your interactions with others work over a variety of time and space constants. Similarly, cells interact with their nearest neighbors and with more distant cells in other tissues: there are fast interactions, but others take time.

This is where the brain is unique: the time and space constants that determine cell interactions in brain vary over orders of magnitude, in a fashion that is simply not reflected in other tissues. Neural cells form complex networks unique to the brain. This means that when tissue is lost to an insult such as a stroke, the impact is not merely local, substantial though that impact might be, but more widely profound. To understand what this means, we need to consider the neural network properties of brain cells.

Brain Networks

Brain cells form an enormous variety of networks, and different brain regions—cerebral cortex, hippocampus, olfactory bulb—each have unique attributes. But all neural networks share some characteristic features. Each, for example, is characterized by the same broad classes of cells (detailed in box 1). One class of

Why Doesn't Brain Repair Work?

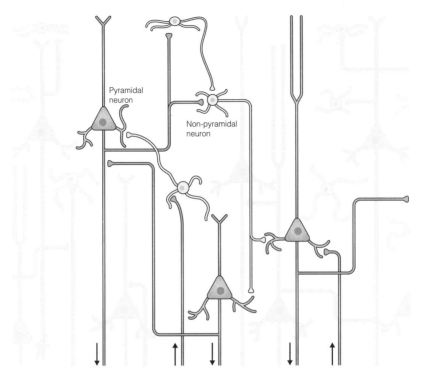

Figure 2.3
Simplified circuit diagram for the cerebral cortex, showing pyramidal neurons, the output neurons of the cortex, and nonpyramidal neurons, the intrinsic inhibitory neurons of the cortex.

neurons is the excitatory, projection neuron. In the cerebral cortex this is the pyramidal cell (figure 2.3). These cells each extend a nerve fiber (an axon) away from the cerebral cortex into another part of the brain, where it connects with other neuron types. Pyramidal neurons have a large range of targets, which is one reason why theirs is a much more disparate class of cells than its unitary name suggests. Different pyramidal cells project down

Box 1
Types of Nerve Cells

Neurons

1. *Projection neurons* are the large nerve cells that connect different brain regions. They provide the major output from any area of the brain, passing information to neighboring regions.

   In the cerebral cortex, the projection neurons are pyramidal neurons, which send connections between cortical areas, and out from the cortex to other regions of the brain or central nervous system, such as the striatum, the cerebellum, and the spinal cord.

2. *Interneurons* are the smaller inhibitory nerve cells that are confined entirely within one brain region. They modulate and refine neural activity before it passes from one region to another.

   In the cerebral cortex, nonpyramidal neurons—the intrinsic inhibitory neurons of the cortex—come in many shapes and sizes. But they are all inhibitory; that is, they limit and shape the activity of the pyramidal neurons and modulate cortical activity.

Glia

1. *Astrocytes*. These star-shaped cells. which outnumber the neurons by perhaps 6 to 1, support the metabolic and electrical activity of the neurons and play a crucial role in cellular defense. They help form the blood-brain barrier, the border that protects the brain from any blood-borne pathogens; they form scar tissue if the brain is damaged; and they nourish the neurons as they pursue their high-energy activities.
2. *Oligodendrocytes* generate myelin, the protective insulator that wraps neuronal processes and allows high-speed, high-fidelity communication between neurons.

Microglia

Distinct from the *macroglia*—the astrocytes and oligodendrocytes—*microglia* are the brain's resident macrophages. These little cells are among the first to react to damage and play a key role in modulating the brain's response to injury.

to the spinal cord, cerebellum, and other brain nuclei; others project across the midline of the brain to the other hemisphere. Yet others project to other cortical areas in their own hemisphere. They all, however, share the property of conducting information from their particular area of the cortex to other target regions.

The second class of neurons are often simply called "nonpyramidal neurons" but a better name for them is "interneurons." Unlike the pyramidal neurons, interneurons are inhibitory—that is, they constrain the activity of their target cells—and they work entirely locally within the cortex. They are also quite diverse. More than twenty different classes of interneurons have been identified in cortical structures. They come in all shapes and sizes and, for reasons that are unclear, seem to have captured some of the more colorful names in neuroscience research. There are "basket cells," "chandelier cells," "double bouquet cells," "Martinotti cells," and many more.

What is the point of inhibitory neurons? Like any stable dynamic system, brain networks require both an accelerator and a brake. By inhibiting the excitable pyramidal neurons, the interneurons are able to stabilize the brain's neural networks and shape their output. Much of the output of the cortex is oscillatory and so requires inhibitory activity to give it shape. The great diversity of interneurons reflects the multiple roles they play in modulating and refining the signals that emerge as the pyramidal neurons process the information that comes to them.

The result then is a complex set of neural networks. Sensory information of many varieties—vision, hearing, touch—feed into the cortex and is processed within individual regions by the pyramidal neurons in concert with the modulating influence of the interneurons. These processed data then pass between cortical regions and ultimately out to other, more distant regions of the central nervous system such as the cerebellum, the basal ganglia, and the spinal cord. Hence the cortex's central role in integrating multiple data streams and initiating most of the major functions we associate with the human brain—language, abstract thought, executive function, and many more.

So let's go back to our stroked brain, and ask what's been lost? The diagram in figure 2.4 identifies three areas of cortex, impacted to a greater or lesser extent by the stroke. Area A is clearly the most seriously affected. This area found itself in the core of the stroke, and has gone through the destruction we discussed earlier. Not only have all the neurons and glial cells disappeared, the entire tissue integrity has been lost. Even blood vessels have gone. There is no longer any tissue structure; all has been replaced by a fluid-filled cyst.

Area B looks to have survived: it is still composed of "eloquent brain," to use the neurosurgeon's poetic expression. It might have been part of the penumbra, but was sufficiently well perfused to have recovered. Nonetheless, if we look carefully, not all is well. Some cells have *not* survived. Perhaps excitotoxicity was particularly uncontrolled in their vicinity, the waves of calcium ions particularly intense. Further, some of the connections to and from this region will have been lost. Area A itself was probably connected to this Area B, so those connections have certainly gone. And perhaps fibers connecting Area B to other regions ran through the destroyed Area A, so those connections are now also lost. The loss of these contacts will provoke a reorganization, as the neural networks accommodate the loss and some function is reestablished.

Beyond this acute effect, the loss of contacts also leads to a progressive cell loss. Like members of all well-functioning social networks, brain cells support one another. The functional integrity of individual neurons depends on communication between

Why Doesn't Brain Repair Work?

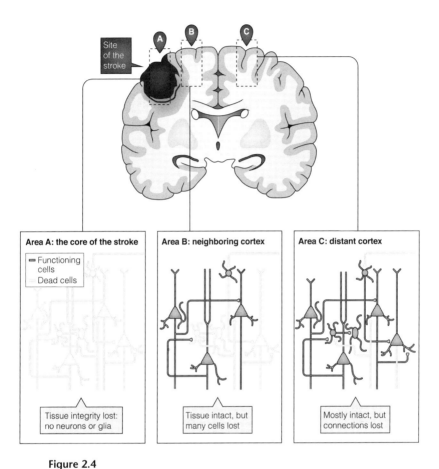

Figure 2.4

Outcome following a stroke. Brain cells in area A of the cortex (stroke core area) are completely lost. Some cells in neighboring area B have survived but others have been lost, as have connections to other areas. And brain cells in more distant area C, though not directly affected, have also lost connections and are functionally disturbed.

them. Activity in one cell reinforces activity in another. So if cells become isolated, they are at risk. Neuroscientists typically distinguish between the fast neurodegeneration that accompanies stroke and the slower neurodegeneration found in disorders such as Alzheimer's disease and Parkinson's disease. But stroke also has a slower component, where cell loss leads to loss of cell connectivity, which in turn leads to further cell death.

What about Area C? Being more distant from the core of the stroke (Area A), cells there had no direct experience of ischemia at all. But like Area B, some cell connections will have been lost, and with them the support that brain cells give to one another. So, even cells in a region far from the core of the stroke will be affected. Neuroscientists have a colorful expression for this phenomenon. "Diaschisis," an almost unpronounceable word, signifies the disturbance in function observed in an area of brain following an insult in a distant brain region. Unique to the brain, diaschisis is a direct result of the neural network properties.

Stem Cells and Cell Replacement

Having seen why a stroke is so destructive, we can appreciate why the brain is so adversely affected. None of this, however, solves our original conundrum: why can't the brain simply replace lost cells like blood or the lining of the gut? Perhaps if we understand how blood and the gut do it, the failure in the brain might become clearer.

Essentially, the answer is simple. Both blood and gut endothelium have stem cells. Stem cells are part of the mechanism some tissues have evolved to maintain stability. This process, called "tissue homeostasis," replenishes the supply of cells, so that tissues have the right cells in the right place at the right time. Stem cells achieve this through two key properties (though, as we'll see later, many so-called stem cells are missing one or the other of these properties). The first is that they are "multipotential," meaning they can generate the full range of cells types that make up a particular tissue. The second is that they are "self-replicative," meaning they can generate more cells like themselves.

Why these two properties are so significant becomes clear when we consider how stem cells replenish their target tissue. Take blood, for example. Blood cells are lost all the time, sometimes as a result of a bleed, but also through their natural turnover. The average life of a red blood cell is 120 days, so on an average day, we need to replace about a hundred billion red blood cells, more if we cut ourselves, menstruate, or donate blood. Where do these blood cells come from? It turns out that we have stem cells living in our bone marrow that generate not only red blood cells, but also all the different types of white blood cells (macrophages, lymphocytes, eosinophils, etc.). These multipotential, self-replicating cells are called "hematopoietic stem cells" (HSCs), and the process of replenishing blood cells is called "hematopoiesis."

We can see immediately why the two key stem cell properties are so important. All blood cell types needs to be maintained: the body can't afford to be without any one of them, so multipotentiality is crucial. And because the blood supply needs to be maintained throughout our lifetimes, our supply of stem cells can't be allowed to run out. So, self-replication is crucial to ensure that the stock of stem cells is continuously replenished.

Hematopoiesis

The process whereby stem cells generate blood is off our primary topic of brain repair, but is worth considering in a little detail for several reasons. HSCs were the first type of mammalian stem cell to be identified (see box 2 for a list of stem cell types discussed

Box 2
Types of Stem Cells

Adult stem cells

Many tissues in the body have stem cells, including skin, gut, and tooth tissues. Listed here are the three adult stem cell types referred to in this book.

1. *Neural stem cells* (NSCs) are defined as multipotential cells that generate neurons and glia. They are found in the adult brain in just a small number of locations, the dentate gyrus being the most important in humans. These adult cells can be grown in culture, but most cultured NSCs used in cell therapies have originated from human fetal sources—aborted human fetuses.

2. *Hematopoietic stem cells* (HSCs) are found in bone marrow and generate blood cells of all types. Rather than doing so directly, however, HSCs give rise to different progenitor cells for each of the different lineages: a common myeloid progenitor to generate red blood cells, megakaryocytes (platelets), and white blood cells such as macrophages, and a common lymphoid progenitor to generate lymphocytes and natural killer cells.

3. *Mesenchymal stem cells* (MSCs) are found in bone marrow as well as in many other tissues such as fat and dental pulp tissues. Also called "mesenchymal stromal cells," these multipotential cells generate a range of connective tissue components: bone, cartilage, and fat. They seem to have therapeutic properties for a range of disorders, but there's considerable uncertainty around their true efficacy and their mode of action.

Pluripotent stem cells

There are two types of pluripotent stem cells discussed in this book.

1. *Embryonic stem cells* (ES cells), derived from the inner cell mass of the embryo, are the true pluripotent cells, which give rise to all the cell types of the human body. ES cells can be directed to

> generate essentially any cell type, from brain and retina cells to liver and heart muscle cells.
>
> 2. *Induced pluripotent stem cells* (iPS cells) can be generated from probably any adult cell by reprogramming, using a combination of factors, typically the four "Yamanaka factors" described in Takahashi and Yamanaka 2006.[2] These artificial cells are functionally equivalent to ES cells, and likewise can give rise to the whole range of cell types.

in this book). Indeed, the pivotal work of pioneers such as James Till and Ernest McCulloch studying bone marrow in Toronto in the 1950s and 1960s really initiated the whole field of stem cell research. Blood stem cells are still by far the most therapeutically important type of stem cell. Perhaps most importantly in the context of our story here, many of the most significant concepts in stem cell biology have emerged from the study of blood, and they define the terminology and mechanistic models that will shape our later discussion of neural stem cells.

As already mentioned, HSCs reside in the bone marrow. These cells are both multipotential and self-replicating as we've already noted, and they are are relatively rare, constituting only about 0.01 percent of bone marrow cells. They also have a property that at first sight that at first sight might seem surprising: they divide very slowly, only once every month or so. For many people, the term "stem cell" evokes images of galloping cell division, a conveyor belt of production to keep up with incessant cellular demand. True stem cells, however, actually divide slowly—one reason why stem cells in the brain were overlooked for so long, as we'll see in a later chapter. This slow rate of division might seem counterintuitive, but it actually serves an important function. The more times a cell divides, the more genetic mutations it accumulates. So keeping the body's cache of stem cells quietly off to one side actually makes good sense.

So how do these lazy stem cells manage to generate billions of blood cell progeny? The real expansion takes place downstream. The HSC generates all the red and white blood cell types, but it generates them *in*directly, through a series of intermediate "progenitor cells," dividing precursor cells that lie between the true stem cells and their downstream progeny. Progenitor cells are a diverse bunch, but let's consider just two types here: "the common myeloid progenitor" and "common lymphoid progenitor." As their names suggest, each progenitor type generates a subset of blood cell types. Those of the myeloid lineage include both the red blood cells and cells of the innate immune system, macrophages and others, and those of the lymphoid lineage, mainly generate lymphocytes, the T-cells and B-cells that have pivotal roles in the immune system. These intermediary progenitors, and those that lie further downstream, are the cells that do the heavy lifting in terms of cell propagation. Under the influence of a number of growth factors—proteins that influence the growth and differentiation of cells—they expand rapidly and generate the huge numbers of cells required in each of these categories.

Stem cells, like good manufacturers everywhere, have to be sensitive to their customers' needs. It won't do for the blood to run out of cells, but nor will it do for cells to build up at any of the intermediate stages. The process requires close coordination. It transpires that biology invented "just-in-time" production methods long before Toyota. For just-in-time to work, the stem cells require timely information on the state of the tissue they are supporting, and fast accurate mechanisms to move the system along or slow it down, as required. These features are built into the "stem cell niche"—a concept now deeply imbedded in stem cell biology—whereby cell production and maintenance is a function of the whole microenvironment in which the stem cells sit, rather than merely the work of the stem cell themselves. The bone marrow niche is truly complex, created at the site of the thick connective tissue lining of the bone itself by an interaction of these connective tissue cells, a particular combination of matrix fibers, growth factors to drive different populations of cells in different directions at different speeds, and the stem cells themselves, communicating directly with one another and modulating the blood cell activity.

To grasp how complex the bone marrow niche is, let's consider just one aspect of that complexity. In one part of the niche, oxygen tension is kept low, so that the hematopoietic stem cells will divide slowly; concurrently, another part of the niche is kept oxygen rich to drive the rapid division of cells in the production line manned by the downstream progenitor cells. So precise is stem cell mediated cell manufacture.

Stem Cell Therapy

The discovery of hematopoietic stem cells has had an enormous impact therapeutically, particularly for blood disorders and cancers. Patients with any one of a number of different leukemias or myelomas are frequently now treated with hematopoietic stem cell therapy. Typically, a patient is irradiated or treated with high-dose chemotherapy to kill the cancerous cells. This treatment also ablates the patient's bone marrow and, with it, their own hematopoietic stem cells. These cells are replaced by donor stem cells injected into the bloodstream, where they make their way to the patient's bone marrow, re-populate the niche, and begin the process of hematopoiesis. Sometimes the transplant is "autologous," that is, composed of the patient's own cells, harvested prior to the irradiation. These have the advantage of being the host's own cells and therefore readily accepted by the patient's immune system. But they carry the risk, depending on the precise nature of the disorder, of reintroducing the cancer. The alternative is to use "allogeneic" cells, harvested from a healthy donor. Here the challenge is getting a sufficiently close immunological match to prevent the patient's immune system from rejecting the cells—or indeed to prevent the engrafted cells from attacking the patient, so-called graft-versus-host disease. In 2012, the number of stem cell donors globally passed the 20 million mark,[3] and the number of applications is still increasing, some even involving the brain.

One of the most interesting recent applications of autologous bone marrow is for multiple sclerosis (MS).[4] Although a disorder of the brain and spinal cord, MS is primarily an autoimmune disease, where, for reasons that are still unclear, the body's immune system starts to attack the myelin sheath, the insulating layer that surrounds most nerve fibers. Inflammation damages this insulation and leaves scarring, which with time can destroy the myelin and damage the nerve fibers themselves. Treatment with bone marrow stem cells seeks to replace the offending lymphoid cells, much as is done in some blood disorders, by replacing them at source with healthy cells. Though still unproven, this might turn out to be one of the first effective stem cell therapies for a brain disorder.

This therapeutic utility underpins much of the success in recent decades in treating blood disorders. Moreover, from a scientific perspective, the result is reassuring. The very same multipotential and self-replicating properties that enable hematopoietic stem cells to replenish the body's blood supply make them effective therapeutically. HMCs can repopulate the entire blood system after it has been destroyed by irradiation. Indeed, this is how these stem cells were discovered in the first place. With advent of the atomic era in the 1950s, radiation sickness became both a clinical and a political issue. Research revealed that the quickly expanding blood progenitors were particularly sensitive to radiation. Hematopoietic stem cells were discovered when researchers transplanted bone marrow into an irradiated mouse, and found colonies of engrafted cells repopulating the bone marrow and the blood. Since these colonies comprised all types of blood progenitors, the founding cell had to have been multipotential. This functional assay was pivotal in identifying and describing stem cells, and stem cell therapy is essentially that same functional assay applied to patients.

Stem cells' therapeutic efficacy is therefore a direct result of their two key properties—of their "stemness." Although this might seem like belaboring an obvious point, it is precisely the point of contention in relation to neural stem cell therapy, as we'll see. Are neural stem cells behaving like other stem cells? Are they repopulating lost cell populations through their ability to rebuild entire systems, or are they doing something else? This is a fundamental question to which we'll return.

Homeostasis versus Repair

If you read much of the stem cell literature—both academic and popular—you could be forgiven for thinking that the primary role of stem cells is tissue repair. But this is a misreading of what stem cells actually do. In fact, their primary role is tissue homeostasis, not repair. This isn't a significant distinction in regard to blood, but it becomes crucial when we consider brain.

In our hematopoietic stem cell example, there is a rapid turnover of blood cells and the stem cells have to maintain those cell populations. This is what is meant by tissue homeostasis in this context.

But in the blood, these stem cells also take care of repair because there is no real distinction to be made between homeostasis and repair. The challenge for the bone marrow niche is not really different whether the blood cells have been removed from circulation by the spleen because they are getting old (homeostasis), or because you've cut your finger (repair). The rate of replacement might be somewhat different, particularly following considerable blood loss, and the hormonal control and the cellular dynamics might vary. But the challenge for the niche is pretty much the same.

The damaged brain is different in two important regards. First, The stroked brain we just encountered has lost more than just cells. The structural integrity of the affected tissue is also lost, destruction exacerbated by the scarring, which of necessity seals off the damaged, but thereby inhibits any potential tissue reconstruction. But, second, there is no brain equivalent of the hematopoietic stem cell. The adult brain has no stem cell with the potential to replace all brain cell types. In chapter 3, we will discuss what types of neural stem cell are to be found in the adult brain, and we'll discover that tissue homeostasis in the brain does involve cell replacement. But unlike the bone marrow, brain tissue homeostasis is not readily adaptable to repair.

Cell Commitment

Before we go back to the brain, there is one more lesson we can learn from the hematopoietic system. This relates to how stem cells know what to do. We've seen how the hematopoietic stem cell gives rise to different populations of progenitor cells, and how each population then generates a subset of blood cell types. Biologists, however, are not satisfied with simply describing what happens. They want to understand *how* it happens. How do the different progenitors arise from the multipotential stem cell? This basic question isn't limited to the hematopoietic system. It's the generic question in developmental biology. How does one thing give rise to another? How does it choose between the alternatives with which it is presented?

Let's call this process "cellular decision making." The conventional way of thinking about this process is that there is a population of equivalent cells—in this case, hematopoietic stem cells. Different individual cells within the population get exposed to different factors, or different concentrations of factors. As a consequence of this exposure, each cell adopts a particular fate; that is, it sets off along a particular trajectory that leads to a particular outcome. For example, the multipotential stem cell would become either a lymphoid, myeloid, or erythroid cell. We could imagine that where the stem cell lay in the niche might determine the magnitude of its exposure to some particular factor, and that factor would then drive its fate in a particular direction.

Consistent with this model, factors have been discovered that drive fate in precisely this fashion. "Erythropoietin" is such a factor, produced by cells in the kidney when they detect low oxygen tension, it acts to increase the production of erthryocytes—red blood cells. Another factor, termed "granulocyte-macrophage colony-stimulating factor" (GM-CSF), acts on hematopoietic stem cells to produce white blood cells of the myeloid variety. GM-CSF is produced by various white blood cells when they are called upon to act, as part of an inflammatory response, for example. So when the body needs more myeloid cells, it calls for more from the bone marrow stem cell niche.

So far, so obvious. But what about the stem cells themselves. How do they respond to those factors? How do they burst into the activity required to generate red or white cells?

The key to this is a group of proteins, collectively called "transcription factors," that switch genes on or off. When erythropoietin or GM-CSF signals to a cell via a receptor on the cell's surface, the receptor in turn sparks off a signal, which cascades into the cell, inducing multiple changes in the cell's shape, energy output, and movement. Crucially, this signaling cascade invades the nucleus of the cell, and starts to work on the genome, turning transcription factors on and off, and so changing gene activity.

Which set of transcription factors gets turned on depends on which growth factor acts on the cell. If the growth factor is erythropoietin, then one transcription factor in particular—*GATA1*—is activated. If the cell is exposed to GM-CSF, then the "PU.1" factor is turned on. Downstream of each of these—*GATA1* and PU.1—is a cascade of other genes, which will in turn direct the behavior of the stem cell. Thus, in response to *GATA1*, the cell will initiate behavior that will lead to the production of red blood cells, whereas PU.1 will orchestrate the production of white blood cells.

So we see an elegant biological mechanism at work. A stem cell listens to its environment, sensitive to the signals coming from cells in other tissues. Then a rapid and effective mechanism drives all its efforts into producing the right kind of progeny. This is tissue homeostasis in action.

Competence

Much of what we know about cell decision making has been known for many years, and has been extrapolated to many other stem cell populations, including those in the nervous system. There is, however, a final element to the mechanism that is only now becoming clear, and again, studies of the bone marrow stem cells are leading the way.

Why Doesn't Brain Repair Work?

These new insights relate to a very old concept, namely, competence. Embryologists have used the term competence for generations to mean the potential of cells to respond to certain stimuli.[5] So, in our example here, hematopoietic stem cells are competent to respond to erythropoietin to generate the erythrocyte lineage, whereas non–stem cells would not be. Competence here means having the cellular machinery to detect the erythropoietin, transduce the signal, and respond appropriately.

But are all stem cells equally competent? In 2008, Hannah Chang and colleagues in Boston published what has become a seminal paper.[6] They were studying mouse hematopoietic progenitor cells, not identical to the human cells we have been discussing here, but broadly similar. They showed that in the mouse stem cell population, there were "high" responders and "low" responders to erythropoietin. Similarly, there were "high" and "low" responders to GM-CSF. The cells that were "high" responders to one factor were "low" responders to the other. Thus, although each cell in the population could respond either way, each was actually primed to go preferentially one way or the other. In fact, each cell was beginning to express the appropriate set of transcription factors for its chosen direction, even before it got the signal and irreversibly committed to adopting that fate.

The advantage of this is clear. The stem cell already has its forces deployed ready for whatever needs arise. It can presumably respond more quickly and effectively. But what is particularly interesting, this is the consequence not of an evolved biological mechanism. Rather it is a consequence of noise.

What's "noise" in this context? A challenge for cellular machinery is that it has to work on a very small scale with very few molecules. The transcription factors we discussed in the previous section, for example, have to get the job done with just a few copies per cell. They have to find their target genes among the twenty thousand or so genes in the genome, while at that subcellular scale, random movements are constantly bashing the various components around. This means noise, which in turn means that different cells in the population are going to vary, even though they are nominally the same. Cellular mechanisms have to be sufficiently robust to function in this environment, but some will be more impacted than others. It transpires that the mechanism described by Hannah Chang and colleagues has a particularly large variance. If measured and compared across cells of the same type, the concentration of most proteins will be seen to vary from cell to cell, but not greatly. But by contrast, a protein called "stem cell antigen 1" (Sca1), which is important in determining competence in this context, varies enormously, more than a thousandfold from one cell to another. And it is stem cells in the Sca1 "high" group that are primed to give rise to myeloid cells, whereas those in the Sca1 "low" group are primed to become erythroid cells.

The beauty of this mechanism for the stem cell is that it is entirely self-regulating. If researchers were to take (as Hannah Chang and her colleagues did) cells from the "low" population and just grow them on their own for some days, they would slowly reestablish the complete "low" to "high" spectrum of cells. Starting with just the "high" population would give the same result. Because the mechanism depended simply on the cellular noise inherent to the system, it would reestablish itself, however much the starting population had been depleted. The whole population has built-in inertia that will always push its cells back to where they started, whatever pressures they have to withstand.

At the start of this chapter, we asked ourselves a question: Why is the brain so poor at repair? We've answered it partway. The damage that follows a stroke could not be readily repaired by simple cell replacement, and, in any case, the brain seems not to have evolved a mechanism for cell replacement equivalent to that of bone marrow stem cells. But we've left questions unanswered. We've noted that there is such a thing as neural stem cells. So, what do they do, and why don't they help out more when the brain gets damaged? Those are questions for the next two chapters.

3 New Cells for Old Brains

If the adult brain doesn't have stem cells equivalent to the blood's HSCs, what stem cells does it have? During the 1990s, research on neural stem cells in mammals was paralysed by controversy. The established dogma, dating back to the great neuroanatomist Santiago Ramón y Cajal and his contemporaries, held that new neurons don't arise in the adult mammalian brain—thus the idea of neural stem cells seemed unnecessary. New findings, however, had suggested that new neurons *do* arise in the adult brain—and that it *does* in fact have neural stem cells. But because, at least in the eyes of some influential neuroscientists, the evidence was not sufficiently convincing to justify the paradigm shift the new findings demanded, an impasse had been reached. Somewhat ironically, the impasse was broken and the dogma laid to rest by an unequivocal finding from research not in mammals, but in birds (as we'll see in the next section).

The basis for the dogma was clear. Nobody had ever identified new neurons, much less neural stem cells, in the adult mammalian brain; and when the adult brain lost neurons as a consequence of damage, they were not replaced. In other words, all the homeostasis and repair functions that bone marrow stem cells

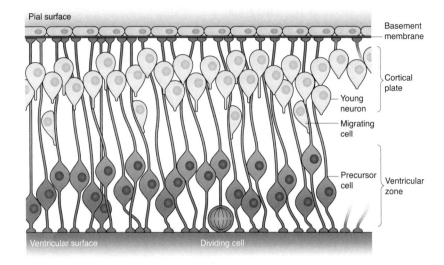

Figure 3.1
Germinal zones in the developing brain. The inner surface, the ventricular zone, is packed with dividing precursor cells. These generate the young neurons that migrate out to form the cortical plate, the gray matter of the developing cortex.

were observed to perform for blood tissue appeared not to exist in adult brain, ergo equivalent stem cells could not exist there either.

As the brain develops in utero, new neurons are generated in great numbers. Germinal regions of the fetal brain are packed with neural progenitor cells (figure 3.1). These divide rapidly, and the new neurons migrate out into the growing brain tissue to begin wiring the complex structures we encountered in the chapter 2. Nonetheless, around the time of birth in most mammals, this neurogenesis dwindles away. The germinal regions start to regress, cell divisions become less frequent, and new neurons no longer emerge. The brain starts to resemble that fixed, compact tangle of cell bodies and cellular processes that characterizes adult neural tissue.

Thus, for decades, there seemed to be no basis on which to postulate the existence of brain stem cells. Moreover, some facts were clear then and remain so now. No one has ever seen a neuron divide. Like heart muscle cells (cardiomyocytes), neurons are postmitotic cells: they lose the ability to divide as they differentiate. So, although some differentiated cells—such as liver hepatocytes—can generate more cells like themselves if required, neurons are denied this mode of propagation. Pockets of germinal cells remain in the young animal, but these are primarily precursors to glia, support cells of the brain and central nervous system, whose production begins later than that of neurons and continues well after birth.

But the "no new neurons in the adult brain" dogma was supported by more than these observations. Many neuroscientists had convinced themselves that the neural circuits of adult mammals are simply too complex to be augmented once they've been established. Simpler nervous systems might get away with it, such as those in species condescendingly labeled "lower vertebrates." Frogs and fishes continue to grow their nervous systems throughout life, but humans, they insisted, do not. Writing in 2006, the neuroscientist Richard Nowakowski argued that:

> The cultural complexity of humans requires not only the constant acquisition of new facts and skills but also the retention of others, most notably language, for many decades, and a stable complement of neurons in the neocortex would seem essential for these abilities.[1]

But just what new neurons can't do that old ones can isn't actually specified in this position. There may well be limits on new neurons that make them incompatible with "the cultural complexity of humans," but what might those limits be? Indeed, what's the neurobiological basis of this complexity, and why does it preclude the addition of new neurons? Without clarity on these issues, no scientifically testable hypotheses can emerge from this position. What experiments could cell scientists perform that would convince a disbeliever that new neurons were really up to the job?

The dogma began to unravel in the 1960s, mainly through the work of Joseph Altman and Gopal Das at MIT.[2] A new method had been invented in the 1950s using a radioactive nucleotide tracer (a nucleotide being one of the components of DNA) to identify precisely when new neurons are generated. When the tracer, called "tritiated thymidine," is injected into a pregnant mammal, it quickly spreads throughout her body, including the placenta and the fetus. It only last in the body for half an hour or so, but during that thirty minutes any dividing cell will take it up and incorporate it into its DNA. This is because dividing cells must replicate their DNA before they divide, so that they have duplicate copies of their genome to pass on to the two daughter cells. Free nucleotides are needed to synthesize the new DNA. So dividing cells, and only dividing cells, will take up this radioactive label.

Having become radiolabeled, the dividing cells in the fetus will pass that label on to their daughter cells, and if those daughter cells themselves then divide, the radiolabel will be shared between the next pairs of daughter cells, and so on. After three or four further rounds of division, the radiolabel will be so diluted as to be undetectable. But if the daughters of that first division never divide, then those daughter cells will retain that radioactive label for as long as they survive in the animal. Neurons are such post-mitotic, non-dividing cells, so newly generated neurons will become indelibly labelled in precisely this way as they pass through their final cell division.

Altman and a number of other scientists had done this experiment: they had injected tritiated thymidine into pregnant rats, mice, and various other mammals, and discovered that many neurons were being generated during the fetal period. In a rat, for example, gestation is about three weeks. Injections during the first 10 to 12 days didn't give rise to very many radiolabeled fetal neurons because no neurons are being formed this early in the rat fetus's development. But later, during the final week of pregnancy, tritiated thymidine injections gave rise to many radiolabeled neurons.

None of this was contentious. The problem arose when Altman and Das injected tritiated thymidine into *adult* animals.[3] The dogma said that it shouldn't be possible to label a neuron with a tracer at this advanced stage: all the neurons should be postmitotic, having been formed during the fetal period, and should therefore be incapable of incorporating the nucleotide tracer. What the two researchers reported, however, was that, even though most neurons couldn't be labeled with the tracer, some could, and that, within the animal's hippocampus, one structure in particular had a large number of labeled neurons. The dentate gyrus is a V-shaped structure that gets its name from its incisor-like shape. It is part of a brain structure called the hippocampal formation.[4] What Altman discovered was that it was home to newly born neurons (figure 3.2).

So where were these new neurons coming from? It seemed to Altman and Das that this hippocampal neurogenesis "might be attributable to the multiplication of small cells with dark nuclei which occur commonly, particularly in young animals, at the

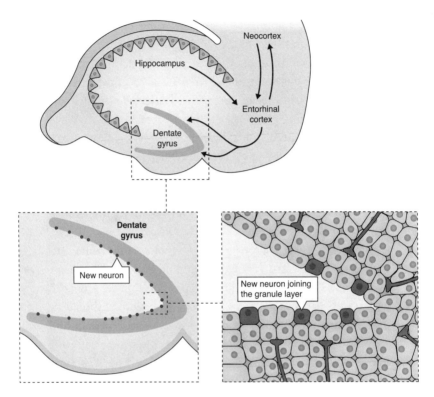

Figure 3.2
Within the hippocampus, a major structure in the mammalian forebrain connecting with other brain regions, lies the V-shaped dentate gyrus with a densely packed layer of granule neurons. On the inner surface of this layer lies a population of neural stem cells that generate new neurons throughout the life of the mammal; these go on to mature and join the granule cell population.

base of the granular layer of the dentate gyrus."[5] Although they didn't use the term, they'd just discovered neural stem cells.

Except that most other neuroscientists didn't believe them, or if they did, they didn't think it was important. Writing in the authoritative *The Neurosciences: Second Study Program* in 1970, Paul Weiss dismissively noted that "sporadic residual straggler neurons have been reported."[6] Other critics went further, denying that the newly formed cells were neurons at all: they must be glial cells, they argued, to be consistent with what everyone knew to be true, that neurons simply don't arise in the adult.

Looking at Altman and Das's 1965 photomicrographs now, what's remarkable is how unequivocal they are. The cells are clearly labeled with the radioactive thymidine, and they look, for all the world to see, like granule neurons. And they're clearly tucked into the underside of the dentate gyrus. That something special is going on in this spot leaps out at you. Even acknowledging that evidence is always clearer in hindsight, how could others have not seen what these two researchers saw?

It took some time to sort out whether a complex mammalian nervous system could truly incorporate new neurons, but the technical issues—were the new cells neurons, and were there enough of them to make a difference?—were quickly resolved. In 2001, Michael Kaplan, a young researcher, first at Tulane University, then at Boston, confirmed the findings of Altman and Das, then added two further refinements.[7] He established the identity of the cells using an electron microscope, which showed, among other things, that the cells had synapses, a pretty definitive proof of their identity. Then he counted the new cells and calculated that, within a three to twelve month period, the number of granule neurons would be increased by half. So the new cells were indeed neurons, and there were enough of them to make a difference.

But the neuroscience community was implacable, and, rather than try to build a scientific career on the basis of his groundbreaking work, Kaplan left research for medicine. At this point, the birds enter the story.

New Neurons in Birds

Fernando Nottebohm is a true biologist. He spent his career studying the behavior of birds, not initially as a biotechnologist might—because of the insight to be gained into human physiology—but because he was intrigued by animal behavior. Sometimes, to the delight of his audience, he would punctuate his research seminars by mimicking the songs of the canaries and zebra finches he was studying. In a fortunate irony of science, one of the most important revelations underpinning mammalian stem cell biology comes from a discovery Nottebohm made about birds.

Songbirds, such as canaries or zebra finches, learn to sing by listening to their fathers' song in the nest. They don't sing that first year, but the next spring, the male birds experience a testosterone surge and start to sing using their fathers' song as a template, enhanced by variations they learn from others, and from practicing their own voice.

This is a sophisticated learned behavior, as complex as anything primates achieve; its neurobiology is correspondingly complex. A pair of motor nuclei, called the "higher vocal center" (HVc) and the "nucleus robustus archistriatalis" (RA), coordinate the patterns of breathing and vocal muscle activity required to produce the song. Concurrently, sensory centers, corresponding to the auditory cortex in mammals, process sound input, and complex regulatory circuits coordinate the sounds the birds hear and the song they sing.

Under the influence of the testosterone, nerve centers in the brains of the male songbirds grow. This is a sexual dimorphism since female brains don't grow in this way. Although this growth had originally been thought to be the result of an increase in the size and complexity of the existing neurons, what became clear from the work of Nottebohm and colleagues was that the number of neurons was increasing and that this was an instance of postnatal neurogenesis. Not only could new neurons be labeled in the higher vocal center of these young birds using the tritiated thymidine protocol employed by Altman, but these new neurons could also be shown to be wired into the motor song circuits.[8] Nottebohm and Paton showed this by injecting a colored dye into the RA region. The injected dye was taken up by the axons of any neurons projecting into that area, then transported back along the axons to the neurons from which they emanate. The two researchers discovered that dye injected into the RA was transported back to neurons in the HVc, and that some of these were exactly those HVc neurons that had been labeled with the tritiated thymidine. This proved that new neurons were not only being formed in the higher vocal center, they were also projecting to their target region, the nucleus robustus archistriatalis. These new projection neurons, generated postnatally, were being integrated into a neural circuit as complex as any found in a primate brain. Neuroscientists began to ask themselves anew: why exactly had we thought new neurons couldn't be incorporated into mature neural circuits?

Adult Neurogenesis Arrives

As the 1990s drew to a close, neural stem cell research was being transformed. New technologies had made neural stem cell studies more robust. A simpler alternative to tritiated thymidine labeling had arisen—a synthetic nucleotide and analog of thymidine called "bromodeoxyuridine" (BrdU)—which, like thymidine, was incorporated into the DNA of dividing cells, but didn't have to be radioactive and was more compatible with modern microscopy. New markers for neurons had also been discovered. Different types of cells are usually identified experimentally by markers—the specific molecules they express. Antibodies can be generated that react specifically with these marker molecules, and the antibodies can be tagged fluorescently, which makes them visible under the microscope. This tagging made the identification of the newly formed cells as neurons much more reliable, and convinced all but the most skeptical neuroscientists that new neurons were indeed being generated in the brains of adult mammals. Microscopy had also found clever ways to improve the imaging of tissue, revealing much that had been overlooked previously.

The biggest changes, however, were within the neuroscience community itself. Stem cell research became respectable. Altman and Kaplan had both struggled to build careers in the face of opposition from a hostile neuroscience establishment. Nottebohm, however, was already an established professor at the Rockefeller University when his pivotal studies emerged. His junior researchers, most notably Arturo Alvarez-Buylla, were able to get good research posts. More neuroscientists were openly declaring an interest in the field.

It turned out that adult neurogenesis in mammals is just as significant as in birds, but in ways that correspond to mammalian brain anatomy and function. The most significant findings were those that followed Altman's original observation. The "small cells with dark nuclei" have now been observed in all mammalian species that have been examined. There is actually a range of stem and progenitor cell types (types 1, 2a, 2b, and 3). These types sit in a niche that has many similarities to the bone marrow niche we discussed in chapter 2, and they seem to represent a developmental progression (whose details are still being clarified) that leads from stem cell through to mature neuron. At the end of this process, the cells eventually express neuronal markers and start to look like young neurons. Many die at this stage, but after two weeks or so, many others, having migrated the short distance up into the granule layer of the dentate gyrus, adopt the morphological and electrophysiological properties of granule neurons. They also receive connections through the "perforant pathway"—the major input into the hippocampus—and dispatch connections to other areas of the hippocampus, just as mature functional granule cells do in the dentate gyrus.

So Altman and Kaplan were both vindicated. We shouldn't be surprised, of course, to see science revealed as a social activity where individuals conform to the view of the majority and dissent is punished. Since the work of Ludwig Fleck in the 1930s and Thomas Kuhn in the 1960s, we have understood that scientific communities tend toward collective thinking, which includes a collective language and method, and which struggles to accommodate dissent. Nonetheless, those of us in the profession still react with disbelief and unease when these cases are unearthed. We think they shouldn't happen and are highly embarrassed when they do. Unfortunately, the cases of Altman and Kaplan won't be the last examples of blind conformity among scientists to appear in this book.

Hippocampal Neurogenesis

So adult neurogenesis is genuine, but if we are to consider neural stem cells and adult-formed neurons to be important, we need to understand their role: what do they do? Well, neuroscientists aren't sure, but at least they're starting to understand what the hippocampus itself does and that seems to involve mood and memory.

The hippocampus has long been a frustrating structure for neuroscientists. Its elaborate sea-horse shape (*hippocampus* is Latin for "sea horse") and its position on the rim of the cerebral cortex suggest an integrative role. Its strong connections to cortex and other parts of the forebrain were consistent with such a core function, but what might that function be? Since the forebrain in evolutionary terms is home to olfaction and the hippocampus is part of the forebrain, much early research focused on the sense of smell. But the absence of direct connections from the hippocampus to the olfactory bulb was discouraging, and that avenue of research gradually dried up. My colleague Jeffrey Grey thought the hippocampus might be involved in anxiety, but no one really quite knew why it was there. W. Maxwell Cowan, a noted hippocampal anatomist, encapsulated the prevailing mood when he titled a lecture "The Hippocampus and the Sense of Frustration."

Nonetheless, the idea had been current for some years that the hippocampus might play a role in memory. This idea had several antecedents, the most significant being the discovery of long-term potentiation (LTP), one of the great leaps forward in the history of neuroscience. In the 1960s, Terje Lømo and Timothy Bliss, working in Oslo, had shown that when the dentate gyrus was stimulated via the perforant pathway, the granule cells of the dentate gyrus duly responded. This was expected, but less expected was the discovery that if this single stimulus was preceded by a long train of stimuli, then the response of the granule cells was enhanced. In other words, the granule cells remembered the previous train of stimuli, and this boosted their subsequent response. This looked then—and still looks now—like a form of memory. Naturally, this discovery attracted attention since the biological basis of memory has always been one of the "big" questions for neuroscientists. It had long been thought that the physical basis of memory might involve alterations at the synapse, certain connections being strengthened and others weakened. LTP was clear evidence of such alterations. So the idea had grown that the hippocampus might have something to do with memory.

Memory is a big subject, and psychologists and neuroscientists have categorized and defined memory in many ways. Some of these categories seem to involve the hippocampus, and some do not. Spatial navigation learning, for example, seems to depend on a functioning hippocampus. A mouse or rat is placed in a "Morris water maze" (named after its inventor, the Edinburgh psychologist Richard Morris), which consists of a big bath of milky water with a platform hidden beneath its surface. The naive rodent swims around until it stumbles upon the platform and pulls itself out. If tested every day, the animal gradually memorizes where the platform is hidden, and after about five or six trials learns to swim there directly. If the animal has a damaged hippocampus, however, it never learns. It takes as long to find the platform after five or six trials, as it did on day one.

Several other forms of memory are similarly dependent on a functioning hippocampus. Fear conditioning, for instance, is the process whereby an animal learns that an otherwise neutral signal, say, a buzzer, in fact signals a distressing stimulus, a foot shock, for example.

If adult-formed neurons really are important for hippocampal function, then you might imagine they're important for these memory functions. This possibility has been examined in experimental animals, and the best that can be said is that the data are inconclusive. Some researchers report that memory is indeed compromised in mice where the formation of new neurons has been prevented; others report that memory is unaffected.[9]

Resolving these conflicting findings is a major research effort in the field of adult neurogenesis. There are several possibilities. Some researchers have used genetic tricks to kill the dividing hippocampal progenitors; others have used drugs or radiation, but all of these approaches have unintended consequences. And how long after the progenitors are killed do the memory deficits reveal themselves? Is it long-term memory researchers should be concentrating on, or short-term? Might the baby-neurons have different roles at different stages of their development? These are all complicated questions.

There's also a potential trap here. We've already witnessed one example of scientists thinking they know the answer in advance, and then being so blinkered they miss important clues. It is important to remember that the newly formed neurons remain in a minority at any one time. Killing just them, while sparing most dentate gyrus neurons, is likely to result in quite subtle outcomes. Important for scientists, therefore, to avoid overstating the importance of neurogenesis, what has been termed "neurogenic evangelism."[10]

Nonetheless, one area of consensus is emerging. New neurons appear to have a role in pattern separation, the ability to distinguish similar stimuli. Learning to distinguish between two very different objects is easy. Most of us learn quickly to distinguish fruit from leaves and to eat the one and not the other. We learn quite quickly to distinguish apples from pears, and develop a preference. But how long might it take us to learn that red apples are sweet, but green apples are sour, particularly if we'd once eaten a sweet green pear?

There's good evidence—both experimental and theoretical—that the dentate gyrus is where this pattern separation occurs. With about five to ten times more neurons than its principal input from the perforant pathway, the dentate gyrus has the capacity to expand incoming signals and to project them onto a higher-dimension space, allowing separation of this input. This specialized anatomy has parallels with the computing techniques used for machine learning. Moreover, experimental evidence suggests that the response characteristics of the dentate gyrus's granule cells are highly tuned: they have a low threshold for becoming activated, responding rapidly to differing inputs, even to subtle inputs that occur at a low intensity or frequency.

Mice with compromised adult neurogenesis have greater difficultly distinguishing similar objects. A mouse might learn, for example, to pick the correct path in a maze by getting a reward for taking it. Then, however, the task is changed, and the animal needs to adapt its response. If the new path is distinctly different from the old, then mice with compromised neurogenesis do as well as unlesioned ones, but the more similar the new path is to the old, the more likely the compromised animals are to make mistakes. Thus new neurons seem to be required to maintain the integrity of this critical cognitive function.

Human Neurogenesis

The hippocampus in humans and in rodents appears pretty similar, making it a fair bet that adult neurogenesis is common to both. Several studies have examined the adult human hippocampus looking for the same populations of stem cells and newly formed neurons that had been observed in the hippocampus of adult rats and mice. Most of these investigations found what they were looking for, cells in humans with the same appearance, carrying the same markers, as the cells seen in rodents. But these results should be treated with caution. Cell markers in particular can't always be relied upon since they can be expressed by different cells in different species. In mice, for example, a protein called "doublecortin" (DCX) is a good marker of young neurons. Cells expressing DCX were indeed discovered in the human dentate gyrus, and they might therefore be assumed to be young neurons also. But since doublecortin may be expressed by a different set of cells in humans, not just by young neurons, such data have to be interpreted carefully.[11]

In a mouse, of course, these cells can be labeled with bromodeoxyuridine to show that they are indeed adult-formed cells, but it would be hard to justify dosing a human with this toxic experimental compound, then doing a postmortem analysis of the dentate gyrus. Or so it would seem at first sight. Thanks primarily to two clever experiments, neuroscientists are now pretty certain that neurogenesis happens in the human dentate gyrus. The first experiment used exactly the same BrdU analysis outlined above. It would be unethical of course to dose a human with such a compound just to evaluate neurogenesis, but fortuitously, it happens that some cancer patients are dosed with BrdU as a diagnostic. This allows oncologists to estimate how much cell division is taking place in their patients' tumors. These being aggressive cancers, some of the patients don't survive very long after receiving the BrdU injections. So having received the patients' prior permission, researchers can examine their brain tissue after death for the incorporation of BrdU into dentate granule cells.

This was done by Peter Eriksson and colleagues in Sweden and California.[12] When they examined this postmortem hippocampal tissue, the researchers discovered adult-formed neurons, just as Joseph Altman had in rats all those years previously. Although Eriksson and colleagues could not confirm that the neurons they observed were physiologically active, or that they really contributed to hippocampal circuitry, the cells had all the right markers and were positioned and shaped exactly like their equivalents in mice.

The second clever experiment used an even quirkier experimental paradigm. In the two decades before the nuclear test ban treaty in 1963, above-ground nuclear bomb testing had led to the release of multiple radioactive isotopes into the air, including carbon 14 (^{14}C). The atmospheric concentration of carbon 14 doubled as a consequence. Through plant photosynthesis, all isotopes of carbon react with oxygen in the air to become carbon dioxide, and become incorporated into the food chain including ultimately into humans, and into their DNA. Jonas Frisén and colleagues at the Karolinska Institute in Stockholm realized that this radioactive label was a means of dating when neurons had been formed, equivalent to the thymidine labeling that Altman had employed. Neurons that became postmitotic before the rise in atmospheric carbon 14 would have less carbon 14 in their DNA than neurons that were generated after the rise. So a person born before the ^{14}C increase—say, in the 1930s—could have neurons with a high carbon 14 concentration only if those neurons had been generated post-1945.

When DNA from brain cells of people born before the ^{14}C increase were analyzed, low levels of carbon 14 were found in most brain regions, consistent with the fact that neurons in most brain regions are generated before birth. In neurons taken from the hippocampus, however, there was more carbon 14 even though these individuals were adults before nuclear testing commenced. So, these hippocampal neurons must have been generated postnatally. In fact, when these data were modeled mathematically, they suggested that neurogenesis is even more marked in the human than in the mouse hippocampus. Quantitative data suggest that about 10 percent of mouse dentate granule neurons are turned over in the lifetime of the mouse, the other 90 percent apparently being generated before the birth of the animal with little subsequent replacement. In humans, however, it seems that there is 100 percent turnover: with time, all human dentate granule neurons will be replaced with adult-formed neurons. So, rather than adult neurogenesis being attenuated in humans, as some neuroscientists had supposed, it appears to be augmented. So much for "the cultural complexity of humans" being incompatible with new neurons.

Though the discussion of adult hippocampal neurogenesis might appear to have taken us somewhat away from our main topic, stem cell repair of the brain, it has helped us lay to rest any suggestion that mammalian brain circuits are too complicated to allow the addition of new neurons. Indeed, if avian vocal motor control and mammalian spatial memory systems can take—indeed require—new neurons, then conceivably such neurons could be incorporated into *any* neural circuit.

This is an important conclusion in relation to brain repair. As we noted in discussing figure 2.4, following stroke damage, whereas some areas of the brain are substantially destroyed, other regions are left structurally intact but with missing neurons. If new neurons could be put in their place, we might expect a functional improvement. Other disorders offer a similar opportunity. Alzheimer's disease, Huntington's disease, and Parkinson's disease all share a common feature—the slow loss of individual neurons from specific brain regions. Again, even limited cell replacement could have a positive therapeutic effect.

But before we get too enthusiastic about this prospect, let's acknowledge a few difficulties. Replacement of lost neurons following stroke damage is likely to be very different from the homeostatic turnover of neurons that neuroscientists see in the dentate gyrus. Moreover, researchers don't yet understand the rules for the addition of new neurons to established neural circuits, even when it occurs naturally. Although the dentate gyrus evolved to include adult neurogenesis, other brain regions did not. This might be an accident of evolution, or there might be good reasons why these circuits don't include new cells.

In truth, neuroscientists currently have a poor understanding of the advantages and disadvantages of adding new neurons to neural circuits. Since, at any point in time, old neurons substantially outnumber young ones in the dentate gyrus, any advantage of having the young cells might be expected to be subtle. Moreover, young neurons have some distinct properties. They have increased electrical excitability (are more likely to fire) and enhanced synaptic plasticity (make new contacts more readily), properties that peak at about four to six weeks after the neurons are generated. The young neurons appear ideally suited to encode new information. Mathematical models indicate that these properties probably enhance the particular pattern separation function of the dentate gyrus.[13] Nonetheless, it's not hard to imagine that, in other circumstances, new neurons might be more disruptive than helpful.

Mood

No sooner had adult hippocampal neurogenesis been identified than it was discovered to vary between animals. Some of the variance was revealed to be the result of different ages. In a number of species, older animals make 5 to 10 times fewer new neurons than younger animals do. This finding raises obvious questions: Why does neurogenesis change? What difference does such change make? We now know that neurogenesis is sensitive to many different factors, and that what changes as a result are memory and mood.

In experimental animals, hippocampal neurogenesis is sensitive to environment, exercise, social isolation, and sleep deprivation. If animals are fed diets that include flavonoids (found in blueberries and chocolate), polyphenols (found in red wine, nuts, and berries) or omega-3 fatty acids (found in oily fish), then neurogenesis is enhanced and spatial memory concomitantly improved. If these positive factors are withdrawn, neurogenesis is reduced, spatial memory deteriorates and there's a negative effect on mood. The experimental animals become depressed, or at least they display behaviors interpreted as equivalent to depression in humans. They show a reduced interest in drinking sugary water (which they otherwise like), and they give up more easily if forced to swim (which they don't like).

Among the most significant factors to impact neurogenesis and mood is stress. Stress reduces neurogenesis and increases anxiety and depression. The biology of stress is complex, but a pivotal mechanism appears to involve cortisol, a hormone secreted by the adrenal gland important in the regulation of an animal's energy supplies and cardiovascular tone. A raised level of cortisol tends to suppress functions the animal can do without, at least temporarily, such as tissue growth and repair, and reproduction. But prolonged excessive cortisol can have negative consequences, including chronic stress and depression. The evidence for this goes back to the groundbreaking studies of Bruce McKewen and Elizabeth Gould, who showed in the 1980s and 1990s that cortisol reduces hippocampal neurogenesis.[14] The mechanism for this is still not entirely clear, but it seems to involve a regulation of a receptor called the "glucocorticoid receptor" (GR) in the hippocampus itself.

The chronic stress and depression associated with raised cortisol levels are linked to the negative impact cortisol has on hippocampal neurogenesis. If experimental animals are exposed to stressful stimuli, their cortisol levels rise, and the same seems to be true for humans. Many patients with major depression have chronically raised cortisol levels. Remarkably, some antidepressant drugs seem to increase hippocampal neurogenesis, particularly when it has been reduced as a result of stress mediated by cortisol. There is convincing evidence now that this might be at least part of the mode of action of these drugs.[15]

Adult Neurogenesis in Other Brain Regions

We've concentrated on hippocampal neurogenesis because it's the most important area of adult neurogenesis in humans. Other species, however, have evolved other priorities. In fact, every vertebrate that's been studied seems to have a different combination of sites with active neurogenesis in the adult.[16] These findings should caution us against any global statements based on our invariably limited knowledge about what is or is not fundamentally possible in biology. If a particular feature is of advantage to a species in the particular environment it occupies and is biologically plausible, then it may well evolve in that species. Mammals have noticeably fewer regions of adult neurogenesis than "lower vertebrates"—typically, two such regions—whereas zebrafish, for example, have sixteen. Hippocampal neurogenesis appears to be recently evolved and particularly prominent in primates. One current theory is that it enhances precisely the type of flexibility that's so characteristic of human cognition.[17] It may prove that, contrary to expectations, the cultural complexity of humans is directly downstream of our capacity to generate new neurons.

The Subependymal Zone

Despite the primacy of neurogenesis in the hippocampus, we need to glance at one other region of the mammalian brain before moving on. Like the dentate gyrus, the subependymal zone (SEZ) is another stem cell niche found in most mammals.[18] Indeed, the SEZ is probably the region of adult neurogenesis in mammals we know the most about since it is particularly prominent in mice, our most popular experimental species. It appears less prominent in humans, but it may turn out to have an important role if stem cell therapies prove successful in clinical medicine.

Deep in the brain are fluid-filled cavities called "ventricles." Surrounding the ventricles of the forebrain lies the subependymal zone stem cell niche. Like their dentate gyrus equivalents, these SEZ stem cells were overlooked for years, simply because they look like something else. The most common cells in the mammalian nervous system aren't neurons, but rather astrocytes, the support cells. These rather unspectacular cells do much of the heavy lifting in the nervous system, supporting and cleaning up after the more glamorous neurons, which control the more exciting information transfer. Glial cell biologists had been trying for several decades to convince other neuroscientists that astrocytes are just as exciting as neurons, when, quite fortuitously, it was discovered that some astrocytes were living double lives. Astrocytes in the subependymal zone were hiding a secret. These "Scarlet Pimpernels" were in fact neural stem cells. Despite their boring appearance, they were in fact performing one of the most exciting tasks in the adult brain—generating new neurons.

It transpires that astrocytes and neural stem cells share a number of cellular properties, and even under the electron microscope they are hard to tell apart. This has led to the suggestion that *all* astrocytes might have the potential to become neural stem cells, and indeed astrocytes can be readily turned into neural progenitor cells using a technology to which we'll return later on.[19]

What are these secret stem cells doing in the subependymal zone? Like their equivalents in the dentate gyrus, they're generating granule neurons. But unlike them, the SEZ granule neurons don't just tuck themselves into the underside of the granule cell layer, as happens in the dentate gyrus. Rather, they have to undertake one of the longest cellular migrations observed in the rodent nervous system. They leave the wall of the ventricle and join a Serengeti-like herd called the "rostral migratory stream" that moves across the brain to inhabit the forward-most structure of the brain, the olfactory bulb. There they become olfactory granule neurons as well as a few other minor neuronal types.

The maintenance of the olfactory bulb is important in rodents. Smell is a major sensory modality for them, and memory associated with olfaction is paramount. Humans are not as dependent on smell, and not surprisingly, the rostral migratory stream is less prominent in our brains. Indeed, the subependymal zone looks quite different in the human forebrain, and there's no consensus yet whether an equivalent population of neural stem cells exists.[20]

So why are we bothering with it? The studies using the ^{14}C labeling that revealed hippocampal neurogenesis in humans also made another novel finding. We've noted already that new neurons were found in the hippocampus, but not in other regions of the human brain such as the cerebral cortex, where adult neurogenesis wasn't expected. But the studies found that another brain region besides the hippocampus—the striatum, in the subcortical basal ganglia of the forebrain—also had new neurons.

The striatum comprises a complex series of circuits that coordinate motor control, motivation, and decision making. It plays a major role in two of the most significant neurodegenerative disorders, Parkinson's disease and Huntington's disease, and we'll consider its function in greater detail when we discuss therapy for these disorders in chapter 4. Because the striatum is also perfused by the middle cerebral artery, it is frequently a major site for stroke damage.

The adult-formed neurons that were discovered in the human striatum were of a particular type. We've seen how there are projection neurons and interneurons in the cerebral cortex. But there are interneurons in the striatum as well, and, like cortical interneurons, the striatal equivalents are varied in size, shape, and function. It turns out that the adult-formed cells of the human striatum are a distinct subset of these interneurons.[21] They express a specific set of signaling molecules, called "calretinin" and "peptide Y." This almost certainly points toward a particular function for these interneurons, but unfortunately not one we yet understand.

Their discovery reinforces the point that cross-species generalizations about adult neurogenesis can be quite wayward. We humans have dispensed with the rostral migratory stream that's so important to rodents, presumably because we don't require their capacity for olfactory memory. Neurogenesis from the subependymal zone hasn't been entirely lost, however. Rather, it seems to have been displaced in primates into the production of the little interneurons that end up on the striatum. It will be very interesting to discover why they're there.

At the end of chapter 2, we posed a pair of questions. We asked: What do neural stem cells do? And why don't they help out more when the brain gets damaged? In this chapter, we've learned that neural stem cells inhabit stem cell niches in the adult brain, and that they're responsible for adult neurogenesis. This has confirmed the important conclusion that at least some adult brain circuits can accommodate new neurons. Nonetheless, we still haven't addressed the question of what they do when the brain gets damaged. Why can't they be like the hematopoietic stem cells and regenerate the whole system?

4   Neural Stem Cells

When Ramón y Cajal wrote that "everything may die, nothing may be regenerated,"[1] he may well have meant it literally, but such an interpretation would not be credible now. Since Ramón y Cajal's time, we've learned that the brain has considerable plasticity, and many cellular elements—dendrites, spines, synapses—are constantly being generated and regenerated. But what about cells? There are stem cells in the brain that produce neurons. How do they behave in response to injury? In chapter 2, we asked: why doesn't brain repair work? Perhaps we first need to ask: just how poor is the brain at repair?

In fact, the stem cell niches of the brain do respond to injury and disease. Studies in experimental animals show that the production and survival of new neurons is increased in response to damage following stroke, for example. One of the earliest such reports came in 2002 from Olle Lindvall and colleagues in Sweden.[2] Using the technology we discussed in chapter 3, they identified newly generated neurons in adult rats that had suffered a stroke. This was induced by blocking a major cerebral artery, mimicking human stroke pathology. Four weeks after the lesion, they observed a 31-fold increase in the number of newly formed neurons detectable in the striatum—the brain area most damaged by the stroke. Moreover, markers confirmed that many of the newly formed cells were of the striatum's major projection neuronal type—medium spiny neurons—not normally produced in the adult striatum.

That doesn't sound so poor: a 31-fold increase sounds like a pretty substantial response. But are we looking at the right statistic? Remember, there is little neurogenesis in the undamaged striatum—just the few interneurons that we discussed in the last chapter. So, a 31-fold increase of not very much is still not very much. In fact, Lindvall and colleagues estimated that "the fraction of dead striatal neurons that has been replaced by the new neurons at 6 weeks after insult is small—only about 0.2 percent."[3] Not quite so impressive.

The researchers also noticed another problem: few of the newly formed neurons survived for very long. The stem cells in the subependymal niche were responding to the damage, but the new neurons they were generating were quickly dying off. Perhaps they were losing their way as they maneuvered out of the niche; perhaps the striatal tissue was just too damaged to accept them; or perhaps they were the wrong kind of neurons.

The Lindvall experiment had the virtue of reproducing a human disorder—stroke—as accurately as possible in an experimental model, but it had the disadvantage that stroke pathology is pretty messy. In both experimental animals and human patients, stroke damage is variable in extent and location, and, as we've already seen in chapter 2, what follows a stroke is pretty close to chaos. What if researchers presented the brain's stem cells with a more precise challenge? What if they took out only a single cell type and left the remaining tissue intact?

This is just what Jeffrey Macklis, Sanjay Magavi, and Blair Leavitt at Harvard engineered.[4] Through a clever manipulation, they managed to kill just pyramidal neurons of the mouse cerebral cortex. They injected fluorescent nanoparticles into the thalamus, a part of the brain with which the pyramidal neurons connect. The nanoparticles were taken up by the terminals of the pyramidal cells and transported back to the cell bodies, so that just pyramidal cells were labeled. Shining a laser onto the surface of the cortex, activated the fluorophores on the nanoparticles, producing a high-energy form of oxygen, which in turn killed the labeled cells. Hence the researchers were able to destroy just the targeted pyramidal cells, leaving the rest of the cortex intact.

As in the 2002 Lindvall experiment, more new neurons were generated than were seen in undamaged cortex. Moreover, Macklis, Magavi, and Leavitt were able to show that these new neurons didn't just look like new pyramidal projection neurons, they acted like them, too, with some actually connecting to the thalamus. But again, the number of these new neurons was tiny. Moreover, other researchers have had trouble replicating the result of the Macklis, Magavi, and Leavitt experiment,[5] so even this reported minimal cell replacement is in doubt.

Of a dog walking on its hind legs, Samuel Johnson famously observed that it might not be well done, but the surprise is to find it done at all. The same might be said of the neurogenic response to brain injury. It's pretty inadequate, but, given the pessimism around cell replacement in the brain, the surprise is to find any replacement at all.

The Right Kind of Neurons

Neuroscientists face a conundrum when doing such experiments. They know that normally, the adult brain produces few neurons of any particular type. They damage the brain, then they see more. But what exactly has happened? They'd like to believe that the stem cell niche has reacted to the damage, acting to restore the lost neurons. This would be the targeted cell replacement that we see in the hematopoietic system. But there's another possibility. Perhaps the niche stem cells are producing the same neurons they usually make— olfactory neurons in the case of the subependymal zone—but these neurons are straying off course because of the damage. Neurons are notoriously promiscuous; they'll try to wire up with the right targets, but if they aren't available, they'll wire up with any that are. (As the old Stephen Stills song goes: "If you can't be with the one you love, honey, love the one you're with.") The result might be that some of the newly formed neurons end up in the space that used to be occupied by the lost cells. Superficially, this might look like cell replacement—the ectopic cells are in the right slots, but like redundant factory workers now stacking supermarket shelves, their hearts just aren't in it.

Thus the first question we need to answer here is: are the niche stem cells really making the right kind of neurons, or are these the wrong kind of neurons turning up in the right place? They say that if it looks like a duck, walks like a duck, and quacks like a duck, then it probably *is* a duck. In principle, it's the same with neurons. If the new neurons can be shown to have the right properties for their replacement role, then it's probably safe to assume they're really doing the job. The problem, however, is that the data are often inadequate: researchers, so to speak, have to decide whether it's a duck, even though they've neither seen it walk nor heard it quack

In the Macklis, Magavi, and Leavitt 2000 experiment, the new neurons really did seem to have gotten it right. Not only did they look like pyramidal neurons, but they also projected to the thalamus as pyramidal neurons should. Yet for all that, there was no evidence of correct physiological activity in this experiment. Which is to say, the new neurons might have achieved everything "duck-like" except "the quack." Appropriate physiological activity would have been difficult to demonstrate. Again, in the 2002 Lindvall experiment, the new neurons looked the part, but it would have been equally difficult to show that they were actually contributing anything useful to neural circuits. In neither experiment would it have been reasonable to have claimed that the new neurons had contributed to brain repair, although I note that, in their 2002 study, Lindvall and colleagues do claim to "provide the first evidence that the adult brain can use neuronal replacement from endogenous precursors to repair itself after stroke."[6]

Researchers mostly rely on markers to help them work out what cells have become. But as we've already seen, these markers can be misleading. Sometimes markers are expressed by more than one cell type. And sometimes they're not cell type markers but maturation markers. Rather than being expressed by just a single cell type, they're expressed on multiple cell types at a particular stage of development. They represent the phase the cells are going through rather than the cells themselves. This phenomenon has been described for some of the markers typically used in the striatum. The 2009 study by Fang Liu and colleagues,[7] for example, suggested that some cells in the damaged striatum might have been misidentified because the markers were not as reliable as was presumed. Moreover, expression of markers might change when tissue gets damaged. Cells might begin to express a marker in response to injury that they never express normally. Finally, of course, markers can't usually be interpreted to mean that the cells are doing anything functional. The marker says that the cells looks like ducks, but it doesn't say they can quack.

All this has put pressure on researchers to devise better means to assess neuronal roles and functions. We'll encounter optogenetics, one such exciting development, in chapter 10 when we consider its application to stem cell therapies, but first let's conclude this discussion by asking whether we can distinguish between any of the alternative explanations for the increased production of new neurons discussed above: is there any direct evidence to be had that neurogenesis can switch on as a consequence of brain damage?

Switching Fate

Probably the cleanest experiment to answer this question was performed by Fiona Doetsch and Constance Scharff in 2001.[8] When they looked at the bird motor control system we considered in chapter 3, they saw, as did we, that there are neurons in the higher vocal center (HVc) nucleus that project to the nucleus robustus archiststriatalis (RA)—which Doetsch and Sharff referred to as the "HVc→RA projection"—and that these neurons undergo normal continuous replacement, just like the granule neurons of the mammalian dentate gyrus. It turns out that there's also another population of HVc neurons that project to a different nucleus called "area X." But the neurons of the HVc→X projection are not normally replaced during adulthood. So the adult stem cell niche of the bird provides the higher vocal center nucleus with new HVc→RA neurons that project to the nucleus robustus archiststriatalis, but not new HVc→X neurons that project to the area X nucleus.

But what happens if either of these two projections gets damaged? Making use of the same clever stratagem that Macklis, Magavi, and Leavitt used to kill off specific populations of neurons in their 2000 experiment, Doetsch and Scharff injected fluorescent nanoparticles into first one then the other of the two target populations, so that either the HVc→RA and the HVc→X populations would each be labeled with the nanoparticles. And, just like the Macklis team, Doetsch and Scharff used a laser to kill either one or the other of these target populations. How did the stem cell niche respond?

The two researchers found that when the HVc→RA neurons were killed, the stem cell niche responded with a considerable increase in its production of new neurons. Though initially the HVc→RA projection was severely compromised by the damage, three months later, it had been restored to normal. By contrast, when the HVc→X neurons were killed, the HVc→X projection never recovered. In other words, responding to brain damage, the stem cell niche could produce more neurons of the type it was already committed to producing, but it couldn't switch to produce neurons of another type.

It's as if the stem cell niche were a ball-bearing factory. If demand for ball bearings goes up, the factory just makes more. But if there's suddenly an increased demand for paperclips, then too bad. Maybe they don't have the equipment to make paperclips, or the expertise. Maybe the marketing department doesn't even recognize that there is an increased demand for paperclips: all they watch is ball-bearing sales. This seems to be how the forebrain stem cell niche operates.

We have to be careful not to extrapolate too far from this single result achieved by Doetsch and Scharff. Nonetheless, retooling appears to be one of the major challenges in the field of neural stem cell therapy. How can neural stem cells be coaxed out of their comfort zone? How can they be made to generate neurons they don't otherwise produce?

The Predicament

So we have a frustrating situation. Demonstrably, there are neural stem cells in the adult brain. We've dispelled the theory that adult neurogenesis is incompatible with complex brain circuits. So there is at least the possibility that if neural stem cells could respond to cell loss—like the bone marrow stem cells do—then some replacement could take place. But like the ball-bearing factory, the niche is too stuck in its ways.

Neuroscientists see two ways forward. One is to work out how to drive the brain stem cells to make the cells required: in other words, to convince the ball-bearing factory to make paperclips. Considerable progress has been made in this direction in recent years, albeit only in experimental animals. It turns out that there are several stem cell types in the brain that might be coaxed into taking on this job, and we'll meet them in chapter 12, when we consider how a patient's own neural stem cells could be repurposed directly to generate new neurons.

The other approach, with the longer heritage, is to bypass the brain's ineffectual stem cells and put the right cells in the right place directly. We'll discuss this shortly, but first let's look a bit more closely at how difficult it can be to define "the right cells in the right place."

Parkinson's Disease and "Dopaminergic Cells"

Some diseases have a pathology that is so iniquitous that one is inclined to imagine that it was diabolically conceived. Parkinson's disease is such a disorder. It is characterized by a slow loss of motor control. In the early stages, the signs might be slight—a tremor, a failure to swing the arms evenly while walking—but this loss increases insidiously. The face freezes into a mask; balance is lost; and walking halts abruptly. Eventually, many sufferers simply get stuck in the "off" state. Many sufferers may be walking across a room when they lose the ability to maintain the movement and are left rooted to the spot.

These symptom are the result of our now familiar adversary, neuronal loss, and the cells in question have the grand title of the dopaminergic neurons of the substantia nigra pars compacta. These nerve cells die, for reasons that scientists are still struggling to clarify, and since they play a pivotal role in maintaining the brain circuits that initiate voluntary motion, a movement disorder is the clinical consequence. But the truly diabolical touch is this: by the time the patient gets a diagnosis, close to half of this neuronal population has already been lost.[9] Even if we could devise a strategy to save afflicted neurons, by the time the disease presents, the patient has probably lost over two hundred thousand dopaminergic neurons. Ironically, as the disease progresses, the rate of neuron loss seems to slow, but, by then of course, it's too late. The damage is done.

If ever there was a need for cell replacement it is here,[10] and this is in fact where cell transplantation into the brain began, at least in recent times.[11] But it began with a mistaken oversimplification that we need to grasp if we're to understand the challenge of cell therapy.

Normally, dopaminergic neurons form a pathway from the substantia nigra to the striatum, a brain region we met in the last chapter (figure 4.1), where they connect with striatal neurons and release dopamine, their neurotransmitter. In one of the most expansive pathways in the brain, between a quarter and

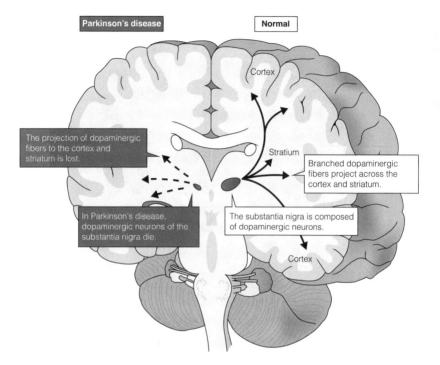

Figure 4.1
Dopamine and Parkinson's disease. Dopaminergic neurons of the substantia nigra normally project widely through the striatum and forebrain (right side of figure). In Parkinson's disease, however, these dopaminergic neurons die, and this projection is lost (left side of figure).

half a million dopaminergic neurons make this projection. This is not a particularly large number in the context of the human brain, but each of these cells forms between one and two million connections with its target striatal cells. The distance from the substantia nigra to the striatum is probably less than ten centimeters, yet each dopaminergic neuron elaborates some four and a half meters of axon; indicating just how much branching and rebranching their processes undergo.[12] This makes these neurons among the most ramifying cells in the whole brain, a neural fountain that showers the striatum in dopamine.

Neural connectivity is normally thought to be very specific. Neurons make precise synapses on their target cells, and deliver their neurotransmitter within a tenth of a micrometer or so of these targets. Moreover, neurons have mechanisms to isolate synapses, so that the transmitter signals don't spread too far from their intended sites of action. But just the density of dopaminergic synapses in the striatum raises the possibility that the dopamine pathway may be different. If cells could be introduced into the striatum that would inundate the tissue with dopamine, even without specific synaptic connections, perhaps this would counter the effect of dopaminergic cell loss.

Existing drug therapy suggested this might work. Patients can't be administered with dopamine directly because its charged structure cannot penetrate the brain. They can, however, receive it *in*directly by mouth, as L-DOPA, a precursor the brain then converts into dopamine. This treatment with L-DOPA been so successful since its introduction in the 1960s, that it has become the frontline treatment for Parkinson's disease.

If just boosting the dopaminergic signal is so effective, what if cells secreting dopamine could be engrafted into the striatum? This was attempted in patients in the 1980s, first in Sweden and later in a number of other countries. Had these researchers used proper dopaminergic neurons for their initial trials, the history of cell therapy might have been different, but, instead, the first cells they used were adrenal medulla cells. Now the adrenal medulla is widely known as the source of adrenaline (epinephrine), the hormone behind the famous "fight or flight" response, whereby the body prepares for action in response to a potential threat. In addition to adrenaline, however, the adrenal medulla also makes dopamine. In fact, these two catecholamines are very closely related. So although transplanting adrenal medulla cells into the brain at first sounds alarming, there's a clear logic to it. If just secreting dopamine into the striatum would be efficacious in treating Parkinson's patients, what does it matter if it comes from dopaminergic neurons or from adrenal medulla cells, or indeed from any old dopamine-producing cells?

The problem arose when researchers failed to realize that while this was a reasonable (if somewhat unconventional) hypothesis, there were lots of reasons why it might not be true. First, the adrenal medullary cells were not brain cells, and might not behave appropriately in that strange environment. Second, even though adrenal medulla cells can indeed become neuron-like in culture, the neurons they resemble most are sympathetic neurons, not striatal dopaminergic neurons, and while they make some dopamine, they make mostly noradrenaline. You might imagine that some serious research would ensue before such a therapy found its way into the clinic.

In fact, the first convincing report of adrenal medulla implants working in rats with experimentally induced parkinsonism appeared in 1981. The first patient treatment with the same technique emerged the following year. Then, through the 1980s, a series of Parkinson's patients—first in Sweden, then in Mexico—were given transplants of their own adrenal medulla tissue engrafted into the striatum. Eventually, several hundred patients were treated with this approach. Though initial reports were positive, even enthusiastic,[13] it subsequently transpired that there was little or no efficacy. There was even a suggestion of unacceptable side effects and increased morbidity and mortality.[14] The most telling observation was that the engrafted tissue didn't even survive for long, so any dopamine replacement could only have been transient. Clearly, the transplant was not doing what was intended, and the therapy was quite rapidly withdrawn..

What did we learn from this episode? First quite simply, we learnt that if we were going to pursue this approach, we needed a better source of dopaminergic cells. The field quickly moved on to proper dopaminergic neurons taken from the substantia nigra itself. This has proven to be a more robust strategy, though also controversial at times. We'll look more closely at this avenue of research in the next chapter.

The second point would seem a fairly obvious one. The logistical underpinnings for the therapy were weak. Little had been done to optimize graft size, location, or viability. Grafts weren't standardized. The surgery had not been optimized. Patient selection had not been thought through very carefully. These points may seem obvious, but cell therapy still struggles to get these basic parameters right. As we'll see when we return to Parkinson's disease in chapter 5, when substantial controlled trials were finally undertaken in the 1990s, they were still subject to this criticism.

For many scientists and their critics, there were more profound lessons that needed to be learned; you shouldn't go to clinical trial until you are absolutely sure you know what you are doing. The preclinical data supporting the medullary approach did not provide a clear neurobiological understanding of how the treatment was working. The supposed mode of action of the cells had not been clearly elucidated. The US Food and Drug Administration (FDA) defines "mode of action" as "the means by which a product achieves an intended therapeutic effect of action."[15] As such, it is the pivotal concept around which any therapy is constructed: what is this treatment doing to the body to bring about a positive change? But the simplicity of the concept hides a quite contentious issue: how much do researchers need to understand about how a potential therapy works before they try it on patients?

A pretty good idea, you might imagine, but that is not necessarily the case. Most treatments, be they conventional drugs or more novel therapeutics, emerge from some sort of screening, usually an animal model, or a cellular or chemical assay. This screening would identify a property of the therapy thought to be of value in treatment. In the Parkinson's example we have been considering, this was an animal model: the dopaminergic neurons on one side of a rat's brain are killed experimentally to mimic the cell loss seen in Parkinson's patients. As a consequence, the rat loses motor coordination on one side of its body, while retaining control on the other. This asymmetry is exacerbated by injecting the rat with a dopaminergic drug such as amphetamine. In response to this stimulus, the rat starts to chase its tail, round and round. If a therapy significantly reduces the frequency of this rotation, it is deemed to have efficacy. You might think this is some way from treating elderly Parkinson's patients, but despite its gross artificiality, it has been enormously useful in developing potential therapies. Significantly, regulators such as the Food and Drug Administration (FDA) in the United States or the Medicines and Healthcare products Regulatory Agency (MHRA) in the United Kingdom consider this an "approved model"; that is, they are inclined to allow potential therapies to enter clinical trials if they have proved robustly effective using this model. But note, the assay tells you little about the mode of action. All you know from this assay is that the turning behavior has been reversed. You don't know how the drug—or the implanted cells—brought about this effect.

You might imagine that regulators would require researchers to have a pretty definitive understanding of the mode of action of a novel therapeutic before allowing it into the clinic. Indeed, you might think that the clinical researchers themselves would want such an understanding. It might come as a surprise therefore to learn that this is not actually the case. Certainly, a regulator will expect some rational explanation for the therapeutic approach, but not usually a confirmed mode of action. Why? Because doing so simply sets the bar too high. Many drugs work through poorly understood mechanisms. The time-honored example, of course, is aspirin, an enormously popular drug through the first half of the twentieth century, long before the discovery in the 1970s of prostaglandins, the signalling molecules that we now think mediate aspirin's effects.

Other examples relevant to our consideration of brain disorders would be the class of antidepressant drugs called "selective serotonin re-uptake inhibitors" (SSRIs). These popular drugs, which include the iconic Prozac, were designed to work in line with the "monoamine hypothesis" of depression, which proposed that depression was associated with a reduced availability of the neurotransmitter, serotonin. The SSRI's were supposed to work by increasing the concentration of serotonin in the synapse, and were licenced as medicines with that as the proposed mode of action. But from the outset there were data that didn't conform with this idea, not least the fact that such a drug should act quickly whereas many depressed patients take weeks to respond. We now know that SSRIs have multiple actions in the brain. Interestingly, one way they act is to stimulate hippocampal neurogenesis. There is evidence now that depression might involve a decrease in the production of new dentate gyrus granule neurons and that antidepressants work by reversing this decrease. So, in this and many other cases, drugs were approved with a mode of action that subsequent research showed to be inaccurate.

Rather than requiring a firm mode of action, researchers and regulators actually start from the position that the therapy should be demonstrably safe and efficacious. There should be sufficient data to suggest both that there is a minimal risk of harm to participants in the first clinical trials. Then there should be robust efficacy data from an approved model (such as the rat model just discussed) to suggest that there is at least a reasonable chance the therapy might work in patients. In practice, this entails a delicate risk-benefit analysis. Is the potential benefit of a new therapy worth the risk inherent in clinical trials? Regarding the mode of action, regulators will usually be satisfied with a credible narrative plus a clear program of research, running concurrently with the trial. But is this practice good enough for distinctly novel therapies such as stem cell therapies? This is a question that becomes particularly acute in regard to the therapies we will consider later, which went to clinical trial at a time when it was quite clear that the mode of action was not cell replacement, and was indeed quite unknown.

Controls

If the mode of action of cellular therapies has been a source of ongoing debate, the issue of controls has been even more troublesome. In conventional drug trials, in addition to the group of patients receiving the new medicine, there would usually be a placebo control group. These patients would be treated precisely the same as those in the drug-treated broup, except that the pill or the injection each received would not contain the active drug substance. The need for this control group is clear: the researchers need to be able to distinguish an actual effect of the drug from the "placebo" effect, the positive effect that sometimes follows simply from receiving clinical attention. Of course, having a placebo control group involves deception: patients don't know whether they're receiving the active medicine or nothing more than a sugar pill. For conventional drug trials, the deception is generally (though not universally) thought to be ethically acceptable. For one thing, all the patients would have given their informed consent to the structure of the trials, and for another, their taking a harmless placebo should put them at no significant risk.

In the case of surgery, particularly neurosurgery, the situation is different. Consider what is involved. The placebo arm of the trial should mimic the "active" arm as closely as possible, so patients would need to be anesthetized, taken into the operating room, have a stereotaxic apparatus attached to their heads, a hole drilled into their skulls, then be injected with fluid. Is this ethical?

In fact, the first cell therapy trials for Parkinson's disease had no control arm. This remains the case today for the planned early-stage trials of stem cell therapies, as indeed for most Phase 1 drug trials. But, ultimately, without randomly controlled trials, new medicines highly unlikely to receive regulatory approval. More importantly, without this control, we can never be sure any efficacy is the consequence of the cells and not a placebo effect driven by the high expectations raised by stem cell therapy in desperate, severely afflicted patients

These ethical and practical concerns will be considered further in a later chapter. But first, let's look at how the dopaminergic approach recovered from this shaky start.

5 The Cell Therapy Approach

In 2001, Curt Freed and colleagues published the results of the first placebo-controlled clinical trial of a cell therapy for a neurodegenerative disease. They had engrafted twenty Parkinson's disease patients with human fetal brain cells, while twenty other patients had undergone sham injections, that is they were anaesthetized, had holes drilled in their skulls, but received no cell injection.

This was a milestone, the largest, neural cell therapy trial ever undertaken, but rather than heralding a breakthrough, it appeared to be the end of the road for cellular therapies in the brain. The trial had "ended in disappointment," reported one pharmacology journal.[1] This new treatment for Parkinson's disease had, reported the *Manchester Guardian*, "gone horribly wrong." In a discussion forum, *Drug Discovery Today* asked "Is there a future for neural grafting?"[2] Even some of the strongest advocates of the approach—Steve Dunnett, Anders Björklund, and Olle Lindvall—published a re-evaluation entitled "Cell therapy in Parkinson's Disease—Stop or Go?"[3]

How had it come to this? Why had brain cell therapy seemingly fallen at the first hurdle? The answer lies in the driving dynamics of biomedical research, human hubris, and the sheer complexity and novelty of what was being attempted.

Following the failure of the adrenal medulla transplants into patients, a number of research groups had learned the lesson that "not any old cell would do." If they wanted to replace lost nigral dopaminergic neurons, they needed to do so with nigral dopaminergic neurons. The dopaminergic neurons of the substantia nigra pars compacta that we met in chapter 4 arise from a part of the midbrain called the "ventral mesencephalon," which lies about halfway back from the front of the skull, hidden under its bigger neighbors, the thalamus and the cerebral cortex. If we are to get some of these cells to put into Parkinson's disease patients, where are they going to come from?

The answer, according to Lindvall, Björklund, and their colleagues in Sweden, was to dissect from aborted human fetuses the tiny piece of tissue that was generating these neurons. As luck would have it, some abortions take place as early as 6 to 8 weeks of gestation, just as the young dopaminergic neurons are starting to appear in the fetal midbrain. Animal experiments suggested that if this little piece of tissue were removed, dissociated into single cells, and injected into the striatum of lesioned rats, then the cells would start to differentiate into the required dopaminergic neurons. They would extend axons that connected with the vacant synaptic space on their striatal target cells—the medium spiny neurons—and they would secrete dopamine into that space. And, indeed, the animals so treated showed a therapeutic response—a reduction in the tail-chasing behavior characteristic of lesioned animals in this model.

The striatum might seem an odd place to put the cells. They were after all midbrain cells. Why not put them in the midbrain? The problem was that while in the fetus they would make connections across the short distance from the midbrain to the striatum, in the adult this distance has grown considerably and now traverses a much more complex terrain. So the thought was: put the cells in the striatum, directly where they need to make connections. This meant of course that the cells would fail to receive the inputs that they would have had in the midbrain itself, and so would not themselves be properly controlled. This was thought to be acceptable compromise. This points to a predicament that complicates all efforts to rebuild brain tissue. Brain is normally built before or soon after birth, not in the adult where the permissive environment no longer exists and where further impediments have arisen.

Despite this limitation, and following a series of encouraging studies in rodents and nonhuman primates, the Swedish group, followed by others in the United Kingdom, United States, and elsewhere, decided the approach was sufficiently refined to try it on patients. There followed a series of trials where midbrain cells were collected from aborted human fetuses and injected into Parkinson's patients, directly into the striatum. This was a cautious approach, both small-scale and open-label. It was very experimental, one or two patients at a time, and without any attempt at placebo control.

The results were variable, but encouraging. Many patients showed the same increase in dopamine synapses that had been seen in the rats. This was monitored by injecting patients with tiny amounts of a radioactive tracer, ^{18}F-fluorodopa. This compound is taken up by dopaminergic nerves and can then be visualized by PET imaging. Patients with the graft had an increased ^{18}F-fluorodopa signal, meaning that the transplanted dopaminergic neurons had done their primary job—survived in the patient brain and made dopaminergic terminals.

More importantly, some of the patients seemed to get better, in some cases considerably better. There are many ways to measure the degree of affliction suffered by Parkinson's patients, and clinicians typically employ a battery of tests that measure the level of tremor as well as difficulties with chewing and swallowing, turning in bed, walking and balance. One simple but effective measure is the proportion of time patients spend in the "off" state. Parkinson's patients can quite simply become too rigid to move. They can be engaged in a simple task such as raising a cup to their lips, when their ability to sustain the action is lost, and they freeze. Some of the patients treated with midbrain cells showed substantial improvements in this measure, reducing by half the amount of time spent in this distressing, debilitating "off" state.

An example is the study published by the Swedish group in 1990.[4] They report a 49-year-old man who prior to the operation had a marked loss in ^{18}F-fluorodopa signal particularly on the left side as a consequence of his parkinsonism, and who spent fully 40–50 percent of his day in the "off" state. He was engrafted with cells just into the left side of his striatum, and during the second month following the operation started to see a marked reduction in rigidity. By three months, his time in the "off" state had fallen to about 20 percent, and this remained stable for the six months following the operation.

Such was the situation in advance of the Freed study: promising outcomes, but small scale and fragmentary. Most importantly, just like the adrenal medulla grafts we considered in the last chapter, the approach was not at all standardized. Consider the variables associated with such a novel approach to therapy. You have to decide how many cells to inject, and where precisely to put them. How should they be handled before they are injected? What should they be suspended in? Should they be all in one place, or spread in multiple sites around the striatum? Plus, the surgery is novel. What is the best way to approach the striatum? This part of the forebrain is on the underside, away from the top of the skull. There is no way to get to it except through other brain tissue, which means collateral damage. So where will that damage do the least harm? Should both sides of the brain be grafted, or should it be restricted to just one side, at least initially? Then, there are the patients themselves. Which patients would be most likely to benefit? Probably the youngest, least afflicted, but how to know? Not all patients have the same parade of symptoms. Which would be the most amenable to treatment? What about immunosuppression? These are somebody else's brain cells (albeit those of an unborn fetus). Do you need to protect against tissue rejection? These are all important questions. If you are to undertake a controlled trial on any sort of scale, you have to have decide on them all, whether you have a sound basis for your answers or not.

But the other side of that coin is: how long can you continue with a patient here and a patient there, before you conclude that you must know, once and for all, whether you have a technology that worked? More than a decade had passed since the first engraftment of midbrain neurons. People were getting impatient. Eventually, someone needed to do a properly controlled study with enough numbers to reach a sensible conclusion. The 2001 Freed study group committed themselves to a set of standardized variables, and went ahead.

And the outcome was disappointing. Overall, the patients that had received the engrafted cells did no better than those who had the sham operation. This was true whether outcomes were measured by the patients' own assessments of their symptoms or by the clinicians' more objective measures. Some patients did appear to improve, particularly the younger ones, and some individual measures, taken in isolation, did show a small improvement—rigidity, for example, was somewhat reduced—but, overall, the trial suggested that the patients who received cell therapy were no better off than those who did not.

In one sense, it was even worse than that. When the ^{18}F-fluorodopa data were analyzed, it became clear that the treated patients did have an increase in the amount of dopamine at neural terminals in the striatum, though whether this should have been sufficient to show a clinical improvement was itself one of the unknowns. This implied that the transplanted cells had made dopamine, and that it was being utilized. One patient was unlucky enough to be killed in a car accident seven months after the operation. A postmortem analysis of her brain confirmed that the transplanted cells had survived, and she had tens of thousands of new dopaminergic cells in her striatum, just as had been hoped. So, in a technical sense, the trial had succeeded: new dopaminergic neurons had been generated. Still, the patients were no better clinically.

There was a further unfortunate wrinkle: the treatment had substantial side effects. Although the surgery itself emerged as relatively safe, after about a year, a number of the treated patients started to suffer from sudden, protracted uncontrolled and uncoordinated movement of their limbs. Such dyskinesias were previously known to arise in Parkinson's patients, sometimes as a consequence of stress and sometimes when patients would overdo their medication, but generally they lasted only a few minutes. Their appearance following cell transplantation was more worrying. These odd movements were lasting for days or even months, even when the patients stopped taking their L-DOPA medication altogether. Freed and his colleagues feared that the growth of the dopaminergic cells was itself out of control, that more terminals were being formed than were needed, that the new wiring had gone haywire.

A second placebo-controlled study reported two years later in 2003, and the findings were similar: no overall improvement in the patients that received the cell grafts, plus dyskinesias.[5] Moreover, again there was the troubling observation that there was an increased ^{18}F-fluorodopa signal—new dopaminergic terminals had been formed—but this hadn't brought about the anticipated clinical bene.

Reevaluation

All this prompted considerable soul searching among researchers, but no consensus on the way forward. Some thought the cell transplantation approach should be suspended, perhaps indefinitely. After all, two well-powered studies had shown no overall benefit in comparison with sham-operated controls, even though more dopaminergic fibers had been formed as indicated by the ^{18}F-fluorodopa and postmortem analyses. For some the approach had quite simply failed. Worse, the uncontrolled movements that patients were suffering meant that proceeding further could be deemed unethical.

Others had a different perspective. Some of the most prominent practitioners in the field had opposed the placebo-controlled trials in the first place. Not because they were opposed in principle—though, as we've seen, there were ethical grounds for such opposition—but rather because the trials were premature. Important variables had not been optimized. Indeed, several procedures hadn't even been tried before they were adopted in the Freed trials. For example, the surgical approach used in Sweden and elsewhere had been to inject cells into the striatum from above, whereas the Freed study approached from an entirely different direction—from the front above the eyes. Did that make a difference? Also, the preparation and number of cells were quite different from those used in Sweden. Again, did that make a difference? A quite basic objection was that the twelve-month period of the trials simply wasn't long enough. Granted, the cells had survived and made dopaminergic connections, as shown by the ^{18}F-fluorodopa signal, but a year might not have been long enough to see clinical benefit. The transplanted cells might not have been able to work that fast. For these researchers, it was much too early to suggest that cell therapy had failed: for them, it hadn't really been tested properly yet. Writing in June 2001, Ole Isacson, Lars Bjorklund, and Rosario Sanchez-Pernaute simply declared: "Parkinson's Disease: Interpretations of Transplantation Study Are Erroneous."[6]

Whichever position you adopted, however, some things were clear. First, there was some good news, the most obvious being that cells could actually be injected into patients' brains without killing them, or even causing them unacceptable damage. Before these trials, it was by no means clear that that was true. In fact, it seemed quite unlikely. The brain is densely packed with neurons connected by fine, delicate processes—axons and dendrites. It was not obvious that new neurons could be forced into this environment without causing considerable trauma. In the event, most patients suffered only minor side-effects as a consequence of the injection itself.

The second positive was there appeared to be "proof-of-concept." For a researcher investigating a novel therapeutic, proof of concept is important. You might believe from your model and your data that a particular approach will bring benefit to patients, but this idea needs to be tested. You need some proof that your overall concept has some value. The fact that some patients—particularly in the open-label trials, but also some in the controlled trials—did appear to show improvement

The Cell Therapy Approach

was taken by many as evidence that the concept had validity. While there was still some way to go, patients could be helped by this approach. This was a judgment that had to be tempered by the fact that there seemed to be no association between the formation of new dopaminergic terminals and the degree of efficacy. Thus, while a "proof of mechanism" was clearly wanting, many across the field took heart from the fact that a proof of concept was not.

Since the publication of the placebo-controlled studies, many researchers have concluded simply that the fetal dopaminergic cells aren't good enough: we needed better cells. There were two problems with these fetal cells that simply couldn't be overcome: one ethical, the other logistical.

The ethical problem is clear. For many people, abortion is unethical, and the use of tissue acquired through abortion is equally unethical. No abortions were ever undertaken specifically to acquire tissue for transplantation: there is total agreement on all sides that that would be wrong. The abortions were all legal under the appropriate jurisdiction, and no woman would have been approached to consider donation of the fetal tissue until her decision to undergo an abortion had been made. Moreover, had the fetal material not been used for cell transplantation, it would have simply been destroyed. Nonetheless, the fact that primary human fetal tissue was being transplanted would never sit easily with some communities and individuals. In the United States particularly, this practice has come under attack, not only by antiabortion groups, but ultimately by a special panel of the US House of Representatives.[7]

The logistical problem is even more restrictive. The tissue used for the transplant is the tiny piece of fetal brain that will generate the dopaminergic neurons. There are myriad problems with using this as a source of cells for transplantation. It is so small that each fetus provides only a fraction of the cells required for every transplant. Precisely how many fetuses-worth of cells are required is one of the issues yet to be properly resolved, but four or five per side has typically been employed.

Then, not every cell in the piece of midbrain is a dopamine neuron-in-waiting. There are other nerve cells—neurons, glia, progenitor cells—as well as nonneural cells such as blood cells and the cells that line blood vessels. All of these go into the gemisch. Moreover, these are foreign cells, which the host will try to reject. And rejection is likely to be more pronounced if the transplant includes immunologically active cells, which is the case for many of these blood cells. We'll discuss this rejection in more detail in chapter 10.

Moreover, some of the other types of neurons in the mix might actually do harm. There's evidence that the later dyskinesias might not have been the result of exuberant dopamine fibers, as Freed and his colleagues originally thought. Rather, it might have resulted from contamination by neurons of a different type. The bit of fetal midbrain that makes dopamine neurons lies right next to the bit that makes serotonin neurons. Though similar to dopamine in many ways, the neurotransmitter serotonin has quite distinct effects in the striatum. If the grafted cells included too many serotonin neurons, this may well contributed to the dyskinesias observed in some patients.[8]

All of this makes this fetal brain cell approach almost impossible to standardize. An ideal cell therapy—indeed, any therapy really—should be precisely formulated. If you get a tablet from your pharmacist, be it an antibiotic, a hormone, or an antidepressant, you expect it to be precisely the same as last one you were prescribed. You expect it to have the same amount of the active ingredient, to have no more than limited impurities. You expect it to have a prescribed shelf-life, and to be proven stable within that storage period. Most particularly, you expect the medicine to have been shown to be efficacious and safe, and you expect the tablet you are given to be identical to those tablets that had been tested for safety and efficacy. One of the main reasons we have medicines regulatory agencies is to make sure that drug manufacture achieve these standards. Fetal brain cells simply cannot be made to fit these requirements.

And finally, of course, there will never be enough. How many hospitals could manage the simultaneous collection of aborted fetal material, the storage and production, the coordinated elective surgery? This is not an approach that could ever treat more than a minute proportion of all Parkinson's patients.

The conclusion from this line of argument is that we have proof of concept, but we are never going to have a viable therapy until we identify a better source of cells. This logic has led in multiple directions, including the genuine stem cell therapeutics that we will discuss in later chapters.

Young Cells into Old Brains

There was another consideration. The Parkinson's field hadn't been standing still while these cell therapy trials had been going ahead. Understanding of the pathology of Parkinson's disease had advanced, and other competing technologies were appearing. Most promising among these was "deep brain stimulation." A device called a "neurostimulator" is implanted in the brain and delivers electrical impulses to particular brain centers. When it was shown to provide relief to some Parkinson's patients, the

FDA had approved deep brain stimulation for treatment in 1997. With this alternative high-tech therapy now available, why should Parkinson's patients undergo questionable and highly invasive cell transplants?

Other insights had also arisen, the most startling coming from postmortem data. In 2007 and 2008, two independent research groups—one in Sweden, the other in the United States—had made the same disturbing discovery.[9] In this discussion of cell therapy we have concentrated on the dopamine problem. The loss of dopamine cells is responsible for Parkinson's most troubling symptoms, and this is what cell therapies have sought to overcome. But for neuroscientists, the more fundamental question has been why do the dopaminergic cells (and indeed other cells in the Parkinson brain) become distressed in the first place? The answer, though still the subject of debate, clearly involves something called "Lewy bodies," protein deposits found in the brains of deceased Parkinson's patients that are characteristic of the disorder and that become particularly prominent in its later stages.

When researchers analyzed the deposits, they found that they were enriched with a particular protein called "alpha-synuclein." The simple hypothesis arose that this deposition of alpha-synuclein was killing the nerve cells. Almost certainly, it is more complicated than that, and in any case, that hypothesis just begs the question of why the alpha-synuclein gets deposited in the first place. Nonetheless, it looked like this deposition was key to the pathology of the disorder.

The unsettling result came when researchers analysed the brains of patients that had received fetal grafts. These were patients that had taken part in the trials that we have been discussing, and who died some years after having received the transplants. In two such brains in two different labs, the researchers found Lewy bodies in the new dopaminergic neurons that had come from the transplant. So not only had the disease continued to progress in the brains of these patients, but it had also gone on to affect the new young neurons the patients had received as transplants.

Why is this so disturbing? The Lewy bodies have always been thought of as a feature of aging. They appear in Parkinson's disease, but also in the brains of patients suffering from other neurodegenerative diseases of the aged, such as certain forms of dementia. Nobody had ever seen Lewy bodies in young neurons, yet these transplanted neurons were just a few years old. They had been transplanted into an old brain, but were themselves fetal brain cells. Even so, the young neurons were demonstrating this definitive characteristic of age. This was distressing because it meant that cells transplanted into patients could themselves be affected by the disease process. More fundamentally, it meant that something in the Parkinson brain was spreading the pathology from cell to cell, and was capable of traumatizing even young cells.

What might this something be? We don't know for sure, but there is a serious suspicion that it involves something called a prion. British readers above a certain age will certainly remember the "mad cow disease" scare of the 1990s. A progressive, fatal neurodegenerative condition known as "Creutzfeldt-Jakob disease" (CJD) was transmitted from animals to humans through the eating of contaminated beef. There was an initial fear that a serious epidemic was upon us. In the event the contagion was well contained, though not before more than 200 people had lost their lives. The infectious agent, we now know, is a protein. Initially, the idea that a protein, rather than a virus or a bacterium, could transmit disease was highly disputed, but eventually confirmed.

The startling fact about the alpha-synuclein story is that many scientists think it too may be a prion. They think that this protein is moving from cell to cell, aggregating to form Lewy bodies, and inducing a progressive Parkinson pathology. This is not a narrative that is universally accepted by any means, but its relevance to our story is clear. You can't put new young neurons into an aged brain and assume they will remain unaffected by the surrounding pathology.

Standardizing Fetal Cells

Despite these difficulties, work has continued to bring fetal cell transplants to the clinic, particularly in Europe. In 2010, TRANSEURO, a consortium funded by the European Commission, was formed to put fetal grafts in Parkinson's disease on a firmer footing. Now underway in Sweden and the UK, twenty patients will be grafted with fetal cells in an open–label trial—so no placebo group—but followed up for three years. The objective is:

> a step-by-step optimization of all technical aspects of the grafting procedure and patient selection and assessment, in order to improve clinical efficiency and consistency, in the absence of troublesome dyskinesias.[10]

The researchers have also attempted to define precisely the criteria that should guide progress of cellular therapies toward the clinic.[11] This comprises a series of questions that researchers should pose themselves, such as: "what is being transplanted, and what is the proposed mechanism of action?" and "what are the preclinical safety and efficacy data supporting the product?" These are not particularly novel or original questions, but their formulation reflects the anxiety among researchers in this field that they have collectively on occasions let events drive them, rather than the other way around.

The TRANSEURO initiative notwithstanding, I don't suppose many neuroscientists now think that fetal grafts represent the real way forward. Quite simply, there are more promising types of cells, most obviously stem cells, a theme we'll pursue in subsequent chapters. But let's note before moving on that other tactics have been employed to restore lost dopaminergic cells. One approach was to create artificial dopaminergic cells. This strategy took various forms. One was to put the cellular machinery required to make dopamine into cells other than neurons. If you chose cells that were easy to grow and expand, then you could make lots of them, all identical and easy to standardize. Then you would just squirt them into the striatum, and hey-ho, they would make the missing dopamine. Our experience with the adrenal cells should make us pretty skeptical regarding such cleverness, but there were pre-clinical data, and it was pursued.

Other nonneurons cells made it as far as clinical trials. "Spheramine" was a therapeutic product composed of cells taken from the eye. There is a thin layer of cells around the outside of the retina called retinal epithelial cells. These cells have a complex function supporting the photoreceptors of the eye, but they also happen to make dopamine. Spheramine comprised these epithelial cells bound to gelatin beads. Injected into the striatum of Parkinson's patients, these cells-on-beads were to replace the missing dopaminergic neurons. After an encouraging initial Phase 1 clinical trial, the Spheramine failed in phase 2 and was withdrawn by Titan Pharmaceuticals, the trial sponsor, in 2008. This may prove to be the final attempt to use cells other than proper nigral dopaminergic neurons in the treatment of Parkinson's disease.

Huntington's Disease

Before we set aside fetal tissue grafts for something a bit more promising, let's note that the Parkinson's studies were paralleled (if less assiduously) by similar approaches to Huntington's disease.

If Parkinson's pathology is devilish, Huntington's is equally diabolical. It is a genetic disorder, caused by a strange genomic malfunction. The crucial gene—the gene that encodes the huntingtin protein—has an odd repeating structure. Glutamate is one of the twenty or so amino acids from which proteins are constructed. At one point in the normal huntingtin protein there is a stretch of about twenty glutamates, one after the other. Glutamate is encoded by the DNA sequence, CAG. So the *huntingtin* gene has a corresponding sequence of CAG repeats (CAGCAGCAG...etc.), encoding this poly-glutamate stretch. This isn't unique. A number of proteins have evolved a similar structure. But it has proved to be a dangerous design: this repeat is susceptible to error when the gene is copied from one cell to another. In some "at risk" individuals, the copying process can stutter, and instead of 20-plus repeats, the number of glutamates can increase to 30, 40, or more. Some Huntington's patients have been reported to have as many as 250 copies.

This strange mutation has profound functional consequences. Up to about 35 repeats, the huntingtin protein appears to function normally, but beyond that, the mutant protein begins to induce profound changes in the brain. Nerve cells die. As in Parkinson's disease, there are many different nerve cells at risk, but again like Parkinson's disease one population carries the brunt of the assault. The focus again is on the corpus striatum, but this time, the concern is not the dopamine projection to the striatum, rather it is one of the principle striatal neurons: the medium spiny cell. We've already met these cells. They are the cells that receive the dopaminergic projection that goes missing in Parkinson's disease. They make up 95 percent of the neurons of the striatum, and as we have already discovered, they play a pivotal role in initiating and controlling movement.

So why does the mutated huntingtin protein cause the cells to die? Nerve cells are social creatures: they depend on the support of their colleagues. Medium spiny neurons receive an input directly from pyramidal neurons of the cerebral cortex. Not only is this cortical connection a pivotal component of the circuit whereby the medium spiny cells control movement, it also provides crucial support for them by secreting a crucial trophic factor, called "brain-derived neurotrophic factor" (BDNF), which tells the striatal cells they're needed by the cortical cells, a signal necessary to sustain the medium spiny neurons.

It transpires that one function of huntingtin is to drive the expression of BDNF, and the mutated huntingtin fails to perform this task properly.[12] Deprived of BDNF, however, the medium spiny neurons die, and a progressive neurodegeneration spreads across multiple brain regions, causing a devastating clinical outcome: abnormalities of movement and mood as well as cognitive imbalances, dementia, and early death.

So, could these medium spiny neurons be replaced the way the dopaminergic neurons were replaced in Parkinson's disease? Research groups in France, Britain, and the United States tried to do exactly that.[13] They used aborted human fetal material, but rather than taking the midbrain used in the Parkinson's studies, they collected the part of the forebrain—the ganglionic eminence—that generates the medium spiny cells. Could these fetal cells replace the lost medium spiny neurons in the brains of Huntington's patients?

The outcomes of these studies mirror those we've already discussed in relation to Parkinson's disease. A significant proportion of patients showed clinical improvements, but these have tended to be short lived, plateauing after two years before slowly reversing. A small number of patients who received grafts have been examined postmortem. They showed that grafted cells had survived and differentiated into medium spiny cells. There was also evidence that the new neurons had wired up correctly, but long-term survival of the graft was an issue. Significantly, neurodegeneration had continued in brain regions other than the striatum. Just as in the Parkinson's studies, the fundamental progression of the disorder was not halted.

Unlike the Parkinson's studies, the Huntington's disease therapy was never tested in a larger scale controlled trial, and the numbers of patients remain small. So while the clinical data for the two disorders are superficially similar, the Huntington's studies remain more rudimentary.

Referring to these fetal transplantation trials in the past tense isn't entirely appropriate. The TRANSEURO trials in Parkinson's continue, and the patients that have received grafts have not gone away. Some are probably alive today only because they received this therapy, while others have lived with distress and disappointment for the same reason. All advanced therapeutics carry with them the intoxication of previous highs and the burden of earlier inadequacies. They are all experiments with people, and even successful experiments tend to lead not to an answer, but to the next experiment.

Fetal transplantation increasingly appears now to be a bridging technology. Not sufficiently powerful itself to impact significantly on medical practice, it has established the arena from which the key elements of cellular therapies for brain disorders—ethical, logistical, scientific—have emerged. As we'll see in chapters 6 and 7, there's no shortage of candidates to pick up the baton.

6 Cell Replacement

The clinical trials in Parkinson's and Huntington's diseases left neuroscientists looking for a better source of cells—cells that would replace lost neurons without the ethical and logistical constraints of human fetal tissue. As that search has progressed, increasingly varied and exotic alternatives have emerged. Of the therapies currently in clinical trial, however, many emerged from that same fetal material. How then were these problems overcome?

As the discovery of neural stem cells became more accepted through the 1980s, they became increasingly the focus of attention as potential therapeutics. Despite the evidence that indigenous neural stem cells respond rather poorly to brain damage, the idea persisted that if more stem cells could be placed at the actual site of damage following a stroke or an injury, then perhaps they would do the job.

But what cells to use? Initially, the only possible source of human neural stem cells appeared to be either adult or fetal human brain. Most researchers plumped for the later—aborted human fetal brain. But while in one sense, this was merely a continuation of fetal cell transplantation, the stem cell therapies that emerged were really quite distinct. Although it started with the material, this approach focused on a different population of cells:

neural progenitor cells. To understand this distinction, we need to look at fetal brain a little bit more closely.

Turning again to the simplified picture in figure 3.1, we can see that fetal brain is actually an epithelium. All neural tissue, brain or spinal cord, derives originally from the primitive skin of the embryo. This "neurectoderm" folds inward to form a tube, which then becomes the brain and spinal cord, but it never loses its fundamental epithelial structure. As we can see in the figure, there is an outside surface, which lies just under the skull, and an inside surface, which surrounds the fluid-filled ventricles, deep in the brain.

Early in its fetal development, the brain is made up predominantly of two types of cells, progenitor cells and young neurons. The progenitor cells are the true epithelial cells. They span the entire structure from outer to inner surfaces. More significantly, they are the starting cells for all the cell types—neurons and glia—that make up the final adult brain.[1] At this stage, the progenitor cells are doing all the work. They are dividing rapidly to make more progenitor cells like themselves so that the brain is slowly inflating like a balloon, but the progenitors are also generating neurons, which migrate outward, where they start to mature and build the gray matter that becomes the adult brain. In a precise order, the neuroepithelial cells will contribute all the different neuronal and glial types that make up this complex structure.

If a piece of young neural tissue is dissociated (as was done for the fetal grafts), then that mix will be primarily composed of progenitors and neurons. Naturally, the cells of blood itself as well as cells from blood vessels and membranes also contribute to the mix, but the cells of these two primary types will predominate.

It is important to be clear from the outset that this is a gross simplification. One fine-grained analysis of fetal midbrain tissue detected twenty-five different types of neural cells in the mix, including five types of radial glial cells (a particular type of neuroepithelial cell) plus four types of progenitors.[2] Working out how these different types relate to one another and how different functions are parsed among them will keep neuroscientists busy for a generation, I suspect, but the general point still stands: most of the cells are either neuroepithelial progenitors or neurons, albeit both come in multiple different flavors.

So here we see the first difference between the fetal transplant and the stem cell approaches. For the fetal grafts, the important cells were the newly formed neurons. Choosing the correct piece of midbrain guaranteed that these included the young dopaminergic neurons that were intended to replace the lost cells in the Parkinson's patients. For the stem cell approach, however, researchers were after the progenitor cells. They wanted dividing cells.

Stem Cells or not Stem Cells?

The second difference between the two approaches is that, to be of use as stem cells, the progenitor cells had to be expanded. But before discussing that, we need to disentangle some nomenclature. In chapter 2, we noted that stem cells have two key properties. They are self-replicative (they can generate more cells like themselves), and they are multipotential (they can generate a range of differentiated cell types). The hematopoietic stem cells of the bone marrow served as an example. These cells generate all the different blood cell types, while ensuring that they also maintain themselves as a population of bone marrow stem cells throughout the entire lifetime of the organism. We've seen subsequently that the neural stem cells of the dentate gyrus and the subependymal zone do something similar in the adult mammalian brain.

But what should we call the fetal neuroepithelial progenitors? They're clearly multipotential, generating the whole range of brain cells: that's their job during fetal development. They also make more cells like themselves, causing the brain to inflate like a balloon, so they are self-replicative But are they true stem cells? Though they are self-replicative, this is not a property they maintain for the lifetime of the organism. Indeed, as the generation of neurons approaches completion around the time of birth, the number of progenitors drops precipitously.

Their self-replication is also limited in a more nuanced sense. The profile of brain cells being generated—neurons and glia—changes with fetal development. Some types of neurons are generated early during fetal life; others later. There's good evidence that this is partly because the progenitors themselves change. Although younger progenitors truly have the potential to generate the whole range of neurons, older progenitors are more restricted in what they generate. Therefore, calling these cells self-replicative is stretching a point somewhat. So, should we call them "stem cells"?

Well, certainly by the 1990s, that's what neuroscientists were calling them in their most prominent publications.[3] But this was probably less a considered scientific decision than the culmination of three factors. First, biologists are notoriously poor at defining their terms. Quite commonly, a paper in biology will include three or four terms all roughly meaning the same thing, without specifying what is meant by each. So what authors take "stem cell" to mean is often not clear. Second, stem cell scientists were only slowly realizing that "stem cells in culture" didn't always correspond to "stem cells in vivo." Often, you could get a cell in culture to behave quite differently from how it was behaving before you ripped it out of the body. Indeed, often this was entirely intended, as we will discover in a moment.

Finally, there was a faint odor of public opinion being manipulated. Stem cells were becoming all the rage in the 1990s. There had been a big boost to the therapeutic use of stem cells when researchers worked out how to purify hematopoietic stem cells. There were reports of stem cells saving lives of patients with incurable diseases. Prominent in the public's consciousness was the treatment of children with severe combined immunodeficiency (SCID), the so-called bubble boy disease. Children with this genetic deficit are fatally unable to combat infection and could only survive in a totally sterile environment, hence they lived in a plastic protective "bubble." But, when treated with their own stem cells, taken from their own umbilical cord and genetically engineered to repair the genetic fault, these babies were able to survive. Claims for therapeutic success with umbilical cord blood stem cells grew exponentially during the 1990s: some real, others fanciful. Stem cells became a topic of public discourse, so it didn't hurt as a researcher if you could claim to be working on stem cells. For the purist, these neural cells were "progenitors" or "precursors," but certainly not stem cells. But rather than spend the rest of the chapter arguing the point, let's note the problem, accept them as stem cells, and move on.

Whatever they called them, researchers had to be able to grow them. We noted earlier that the potential of stem cells is ephemeral. Once put into tissue culture, they quickly differentiate, and, within a short time, there are no stem cells left. For fetal brain stem cells to be useful, this limitation had to be overcome. To use the jargon, the technology had to be "scalable."

The Problem of Cell Expansion

Essentially, there are two tricks to making your cells divide: you can either treat them with growth factors, or you can manipulate their gene expression to keep them in the cell cycle. Both have been employed to generate scalable stem cell populations for therapeutic use. We'll meet the second of these in more detail in subsequent chapters, but we need to start first with the growth factor approach.

As long ago as 1992, Brent Reynolds and Samuel Weiss, working in Calgary, Canada, reported that they could take neural progenitor cells—such as those found in the fetal brain, or in the stem cell niches of the adult brain—and culture them for extended periods, without them losing their "stemness," their self-replication and multipotency.[4] The trick was to grow them as "neurospheres," balls of cells that stuck to each other but not to any plastic tissue-culture dish, and to treat them with high levels of two growth factors, "epidermal growth factor" (EGF) and "fibroblast growth factor" (FGF). The result, they reported, was a perpetual culture of neural stem cells.

Over the following decade or so, neurosphere technology became something of a cottage industry. It was quickly extended across an wide variety of neural stem cell types: from forebrain to retina, from fetal to adult, from mouse cells to human cells. If you could find a source of neural stem cells, more than than likely you could grow them as neurospheres. The most freakish extreme was probably the narrative describing the work of a Shanghai neurologist, Zhu Jianhong.[5] He was called upon to treat a female patient whose partner at dinner had terminated an argument by plunging a chopstick through her eye socket and into her frontal cortex. The enterprising Dr. Zhu had realized as he removed the chopstick that as well as the remains of lunch, he had recovered a small amount of brain tissue. Cultured appropriately, he had indeed been able to generate neurospheres from these fragments of adult human brain as he described in the *New England Journal of Medicine*. In fact, he reported 16 patients with open brain injuries from which he had been able to recover neural stem cells with this technique.[6]

These, along with more conventionally sourced neural stem cells, found their way first into preclinical studies then increasingly into clinical trials. A significant milestone was the report in 2006 that children suffering from Batten disease, a rare genetic neurodegenerative disorder, had been engrafted with HuCNS-SC® cells (human central nervous system stem cells for neurological disorders). Children who inherit the Batten mutation have a fault in the system that clears away cellular waste, and as a consequence, undergo progressive neurodegeneration, suffering from seizures and blindness; they lose the ability to walk, to talk, and invariably fail to survive past their teens. The cells used for the 2006 study were essentially neurosphere-expanded stem cells that had also been purified using other cell-sorting technology. Six children received these cell grafts, described by *Scientific American* as the "World's first neural stem cell transplant."[7] In a 2009 study, and in a follow-up study in 2013, neuroscientists at StemCells, Inc., the studies' sponsor, were able to present good safety data. Like the Parkinson's patients before them, the treated Batten patients were able to accommodate the transplant injections into their brains with minimal adverse side effects. Three of the six survived for the four years of the follow-up study, a rate described as "consistent with the natural history of the disease," meaning neither better nor worse than would be expected for untreated Batten children. In a decision that must have been devastating for other sufferers and their families, StemCells decided in 2011 not to pursue the therapy further, citing difficulties with patient recruitment as the primary reason.

There were bigger fish to fry. Batten disease is rare, affecting perhaps as few as 150 people in the UK at any one time. By contrast, 40,000 people there live with spinal cord injury, and more than half a million suffer from age-related macular degeneration (ARMD). It turns out that neural stem cell therapy might be effective against both these disorders.

Spinal Cord Therapies

Chapter 1 outlined the neurodegenerative changes that accompany a stroke and suggested that much of that pathology was common to many disorders of the central nervous system. This includes the spinal cord, but in addition injuries here present a uniquely difficult problem.

Spinal cord damage occurs in two waves, first in young men (predominantly) in their teens and twenties as a result of motor vehicle collisions or—in countries with a gun culture, like the US and Brazil—through gunshot wounds. A second phase follows in the elderly as a consequence of falls. The spinal column is depressed or punctured. This squeezes or severs the spinal cord, and induces a set of changes similar to those following stroke—hemorrhage, edema, ischemia, neurodegeneration. Tissue is lost in a fashion with which we are now familiar.

But the unique problem with spinal cord injuries is this. The damage sweeps away not only the tissue directly in the path of the injury, but also the axons of neurons that pass through the area of destruction. The spinal cord contains the neural circuits that control voluntary and involuntary movement throughout

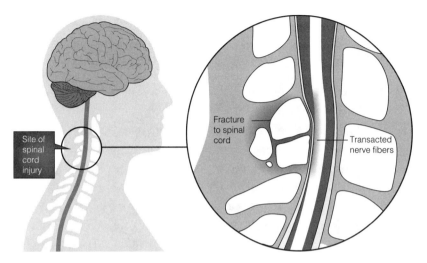

Figure 6.1
Spinal cord injury. As shown here in the neck region, a fracture to the spinal cord compresses the spinal column, and transects nerve fibers running up and down the spinal cord.

the body: movements involved in breathing controlled at neck and chest level; those of legs, bowels and bladder at lower levels. But these spinal circuits are controlled in turn by axons passing from the brain down the entire length of the spinal cord, and if the cord is transected at any specific level, then all the circuits below that level are robbed of their input. So if the break occurs high up in the neck (figure 6.1)—as happened to the actor Christopher Reeve—then all control below that level is lost

Consequently, the case for stem cell repair of the primary tissue loss following spinal cord injury appears similar to the one we posed for stroke earlier: could the lost tissue be regenerated? But this is trumped by a more significant question, one that at first sight seems easier to address: can we induce the severed axons to grow though the area of damage, and re-innervate their targets further down the cord? Not so much a case of replace, more a case of rewire. Of course, rewiring is also a component of the challenge in stroke and Parkinson's disease: any new neurons have to be wired in to work. But in spinal cord repair, it really is the primary objective.

Multiple pathways connect the brain with the spinal cord, and they travel both up and down the spine. Nonetheless, we can get a good flavor of the problem if we focus on the corticospinal tract. This is the most important motor pathway in the human brain and it runs a long way. Its axons originate in the large pyramidal neurons in the motor cortex, cells we encountered in our discussion of cortical function in chapter 2. The axons project backward through the entire extent of the midbrain and hindbrain, then down the length of the spinal cord. They are indeed the longest axons in the human nervous system, some of them running from just below the scalp to near where our tail would attach, if we had one.

The problem of rewiring the corticospinal tract has three components. The first impediment is that following the lesion, a scar forms presenting a barrier to the regrowth of axons. Second, some of the terrain through which the axons need to pass inhibits their growth. A particular problem is myelin. This is the insulating material that surrounds nerve axons, and speeds electrical transmission along the axon. Ironically, this material aids axonal function, but inhibits axons if they have to regrow. The third problem is that axons themselves, once mature, do not have the facility for growth that they had while young.

How might stem cells help? Well, first, of course, they might regenerate new cells to replace the cells lost to injury. Needless to say, that was the original idea, as with much that we've considered so far in this book. But a second possibility also arose (plus a third that we will discuss in chapter 10), namely that the stem cells—or derivatives of the stem cells—might form a bridge. Perhaps they could act as a pathway, or lay a substrate, or somehow metabolize the tissue, in a way that would overcome the blockage and encourage axon growth. In fact, there's evidence for both modes of action.

A number of research groups have now shown that the engraftment of cells from neurospheres can bring about recovery in animals with spinal cord injuries. In a pivotal 2005 study by Brian Cummings and colleagues in California,[8] mice received a controlled blow to the spinal column at midthoracic level. Nine days later, they were engrafted with the neural stem cells. Over the following weeks, the engrafted animals showed improved motor function in comparison with lesioned animals that didn't receive the cells. It was a modest effect, but, generally, the mice with the cells could better coordinate the movement of their fore- and hindlimbs, a function that had been compromised by the spinal cord injury. Moreover, the researchers discovered that the engrafted cells had indeed generated neurons and other neural cells. They seemed to have started to rebuild spinal tissue.

This was an important milestone: human neural stem cells showing efficacy following major spinal cord injury. But the result was influential for a second reason. We noted in chapter 4 when discussing the 2000 Macklis, Magavi, and Leavitt experiment how difficult it can be for experimenters to interpret findings like these. Cummings and colleagues had a similar problem here. The animals' motor performance had improved, and new neurons had appeared. But how to be sure that the improved functional outcome was a consequence of the formation of the new neurons and not of something else altogether?

The scientific way to answer this question would be to take away the new neurons and see if the effect survived. If it did not, then the effect clearly required the presence of the new neurons.

But how to do this? Obviously, the new neurons could not just be cut out. That would have caused further damage and messed up the experiment entirely. The Cummings group came up with a clever solution. Human neurons are much more sensitive to a particular toxin—the diphtheria toxin—than mouse cells. By treating the mice with a controlled dose of this toxin, the experimenters were able to kill all the human neurons derived from the engrafted cells, while leaving the indigenous mouse cells unaffected. And when they did this, the functional improvement disappeared. This was enormously reassuring: these new cells were truly responsible for the improvement. Moreover, subsequent studies suggested that engraftment was not affecting either scar formation or lesion volume. It looked like the formation of the new cells—the neurons or the glial cells—was mediating the recovery.

Armed with the good safety data from the Batten disease trials and these encouraging animal studies, clinical trials were undertaken. In fact, two trials took place. The first involving twelve spinal cord injury patients was completed in 2015; the second involving five patients was terminated prematurely in 2016. Although the early reports suggested efficacy, further studies were abandoned after the early termination of the second study. StemCells, Inc., shut down operations following the announcement of this failure.

So what went wrong? Some thought nothing much was amiss. "It's the nature of science that not every experiment works, and the nature of business is that not every company succeeds." So thought Kevin McCormack of the California Institute of Regenerative Medicine, which had invested substantially in StemCells, Inc.[9]

But others were less sanguine. In their commentary on the failure, Edwin Monuki and colleagues identified a number of issues, and they have a familiar ring. The animal models might not have been appropriate. Important variables, such as the appropriate time following injury, had not been sufficiently investigated. There was particular concern regarding the variability from batch to batch between the cells used for the preclinical work and those used in the clinical trials themselves. There were similar arguments following the Parkinson's trials, and these issues have not gone away.

A comparable trial has also taken place using spinal cord stem cells in another motor disorder, amyotrophic lateral sclerosis (ALS), also known as "motor neuron disease" and more commonly in the United States as "Lou Gehrig's disease." ALS is a particularly devastating neurological disorder. Patients typically survive between two and five years following diagnosis, and the only current treatment, "riluzol," extends life by only a few months. The disorder involves the loss of not just the motor neurons in the spinal cord that drive muscle action, but also the pyramidal neurons that project from the cerebral cortex and enable voluntary control of muscle activity. The disorder is inexorably progressive, with muscle weakness leading to muscle wasting and paralysis of the limbs and the body. Ultimately, vital functions such as breathing and swallowing are affected, eventually leading to death.

Since 2009, the US company Neuralstem, Inc., has been running a trial of neural stem cells as a therapy for ALS, using the human neural stem cell line "NSI-566RSC," derived from human fetal spinal cord, grown in culture, and treated with FGF. The line was expanded and frozen, and this stock has been used for the clinical studies. The line demonstrates clear spinal cord properties in culture, where it can differentiate to give motor neurons, though the cells used for the clinical trial are the undifferentiated stem cells. This structure leaves open the possibility that the cells might differentiate in vivo following grafting, true tissue replacement that might be detectable by imaging. They also might have efficacy in other disorders we have considered, including stroke, for example.

The trial has a complex "risk escalation" structure with twelve patients, grouped into three cohorts. The first were nonambulatory patients, who received transplants into the lumbar region of the spinal cord. The second were less-afflicted ambulatory patients, who again received lumbar grafts. The third group received grafts of cells into the neck region of the spinal cord. A recent report on the study indicated that the protocol appeared safe and well tolerated. There was no evidence that disease progression had been halted, though this small trial was not planned to demonstrate efficacy. Further clinical trials are planned.[10]

Retinal Disorders

It might seem amiss to include retinal therapies in a book about central nervous system repair, but the retina is, of course, an outgrowth of the brain, composed of neural tissue. During development, two eye cups extend from the forebrain but remain connected by two stalks. The stalks become the optic nerves, and the cups become the retinas. Each cup starts in the fetus as a hollow sphere of cells, like a tennis ball, which becomes pushed into a cup shape. Imagine pushing your fingers into the ball such that the two surfaces come together. The front surface—where your fingers had been pushing—is now the inside of the cup, and becomes retina. The back surface, which now forms the outside of the cup, is the pigment epithelium.

Healthy vision requires an intimate interaction between these two layers. The photoreceptors of the retina—the light detecting cells—lie at the interface of the two layers. They are a strangely vulnerable group of cells, in part because the act of turning photons into neural signals is so biochemically stressful that the outer segment of the photoreceptors—the light-sensitive part—is continuously degenerating. The photoreceptors only manage to maintain their integrity because they are nourished and supported by the pigment epithelium.

In a number of sight-threatening disorders, this relationship breaks down (figure 6.2). In "retinitis pigmentosa," inherited

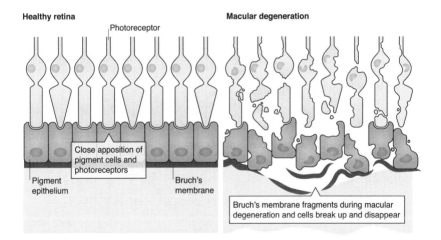

Figure 6.2
Structure of retina. In the healthy retina (left panel), close apposition and interaction of the photoreceptors (rods and cones) and the pigment epithelium are critical for the health and survival of the photoreceptors. In macular degeneration (right panel), Bruch's membrane (on which the pigment cells sit) degenerates, leading to the death of pigment epithelium cells and the degeneration of the photoreceptors.

mutations cause the death of the pigment cells. In the more common "age-related macular degeneration" (ARMD), the center of the retina becomes chronically inflamed resulting in a loss of the pigmented cells and compromised photoreceptors. In both cases, the the prediction would be that replacement of the lost pigmented cells could lead to repair, and just as with brain damage, this hope has driven a search for the right type of cells to achieve this replacement.

In several ways, pigment cell replacement should be easier than the brain cell replacement we've been considering up to this point. First, the space behind the retina is relatively accessible. Injecting material into the eyeball isn't for the squeamish, but ophthalmic surgeons are pretty adept at targeting the virtual space between the retina and the pigment epithelium while doing minimal damage. Second, the obvious replacement cell type—the retinal pigment epithelial cells themselves—can be sourced, most readily from the patient's own retina. Surgeons can detach a piece of pigment epithelium from the periphery of the retina and transplant it to the central region. Of course, this creates some loss of peripheral vision, but this is a price many patients would be willing to pay for improved central vision, the ability to see clearly the object of focus. An alternative source is cadaveric retina, the transplant of a piece of pigment epithelium from a deceased donor. This has also been trialled.

As in spinal cord injury, there are animal models of these disorders. There is, for example, the Royal College of Surgeon's (RCS) Rat, which loses photoreceptors and becomes blind because of a mutation in a gene, MERTK, analogous to the pathology in some human patients. Preclinical studies suggest that pigment epithelium grafts can bring about functional improvement in such animals.

Several groups have now reported results following clinical trials of pigment epithelium in patients.[11] The numbers are small—nine patients in one study, seven in another—and overall, the best one can say about the clinical outcomes is that efficacy is modest: small improvements at best, some patients "stabilized" rather than improved. Why were the outcomes no better? Possibly the photoreceptors were already too damaged to respond. Possibly the blood supply to the graft was inadequate. Perhaps the surgery actually damaged the photoreceptors. Certainly, in some cases the injection caused the retina to detach from the back of the eye. One possibility, of course, is that the patient's own pigment cells were too compromised across the whole retina to be of any real use, so again, we may be looking for a better supply of cells.

The Status of Cell Replacement

In pulling together these two modes of cell replacement, spinal cord cells and retinal pigment epithelium, I've taken some liberties. First, I've ignored other disorders where stem cell studies are ongoing, such as multiple sclerosis and traumatic brain injury, because treatments for retinal and spinal cord damage are probably the two arenas where neural transplants have shown the most promise (as opposed to transplants of other cell types), and where the breakthroughs in cell replacement were arguably most expected. Most importantly, I think, here is where we can draw the clearest lessons, and best see the way forward.

The first question to ask is whether we think, given these relative failures, that the "cell replacement" initiative is dead. Proponents are quite clear that it is not.

> It is the nature of science that not every experiment will work yet even in failure we can learn a lot, and it's our hope that the lessons

> learned...will help inform other researchers and ultimately lead to effective therapies,

concluded Lawrence Goldstein, director of the Stem Cell Program at the University of California San Diego, following StemCells, Inc.'s demise. So what are those lessons?

Clearly, the problems that beset the Parkinson's studies are generic. The step from animal studies to human trials is fraught. The models only approximate generally to the human condition and give little help in sorting out the many logistics: how many cells should be injected, where precisely should they be put, how to assure that different batches of cells have similar equivalent properties. These issues are particularly acute where the animal studies have suggested only marginal outcomes. Researchers driven by commercial concerns—such as those of StemCells, Inc.—want to get on and try their products in clinical trials. They have cells ready to go, and patients desperate to receive them. Others ask: how can you proceed when it is all so experimental? That conundrum has not disappeared.

Two particular concerns were amplified by the StemCells, Inc. episode and were particularly highlighted by their academic collaborators. The first was whether communication had been adequate between researchers, regulators, and patients. Not all the primary data generated from the animal studies were in the public domain before clinical trials commenced. Thus neither the regulators nor patients had full access to the facts when they were deciding whether to proceed. In 2016, Aileen Anderson and Brian Cummings questioned whether patients could be considered to have given truly informed consent if they were not in possession of the complete facts.[12]

This isn't just a theoretical point. Preclinical studies had reported a tendency of the neural stem cells to form "ectopic clusters" following their engraftment in animals' brains. This was not deemed by StemCells, Inc., to pose a serious risk to patients, but should that information have been available to them and their advisors?

A company's research data are, of course, confidential, but how much should they be ethically and legally required to release? It has been pointed out that, even when complete, the findings from clinical trials are often not published. Results from fewer than half of stem cell trials worldwide are eventually published, and fewer still have results entered on the ClinicalTrials.gov website, a database established in 2008 by the US National Institutes of Health (NIH) as a repository of clinical trials data.[13] Though the database is open to researchers worldwide, and though trials funded by the NIH are required to be posted, publishing the data in this way is not obligatory for most studies. Many commentators think it should be, and that failing to do so is unethical.[14] Should StemCells, Inc.'s funders, which included the California Institute of Regenerative Medicine, have insisted on publication? Patients have volunteered for trials. If the trial has a placebo arm, they will have volunteered knowing that they may have to go through surgery without receiving the cells. Surely, they deserve to see the data published, to know that the risk they've undertaken can benefit other patients after them? Certainly, the International Society for Stem Cell Research (ISSCR) guidelines[15] require that data should be disclosed in full, and Larry Goldstein's lessons cannot possibly be learned if the data remain secret. Notably, the results of neither spinal cord trial undertaken by StemCells, Inc., have been posted on ClinicalTrials.gov as of June 2017.

So is cell replacement dead? Well, no, but it has moved on. The London Project to Cure Blindness is now pursuing the pluripotent cell strategy for treatment of age-related macular degeneration, as are researchers in Japan, and at Advanced Cell Technology in the United States. Researchers at Asterias Biotherapeutics in the United States and Keio University in Tokyo are pursuing treatment of spinal cord injury with similarly derived cells. Parkinson's disease researchers have transferred their attention in a similar direction. This "new" type of stem cell now looks to have more promise, and and this is the subject of chapter 10. In the meantime, stem cells have been shown to have some unexpected properties, which in turn have opened up some unexpected for brain repair, and that is the subject of the next chapter.

7 Stem Cells Redefined

In chapter 1, we noted a remarkable observation. Stem cells are injected into the brains of experimental animals damaged by a stroke or some other form of injury. Over the subsequent weeks, the animals show substantial functional improvement, but there is no cell replacement. So what's going on here? How can stem cells be bringing about functional recovery without doing what they were asked to do—replace lost cells? And how should we view this finding? Is it good; despite its serendipity can this approach lead us towards a radical new therapy: or is it bad, convincing vulnerable patients that stem cells really are magic but offering nothing of substance?

One thing is certain: this finding now underpins the most progressed therapies in the field of cell-mediated brain repair, and the outcomes of their advanced clinical trials will determine to a considerable extent the future of the field as a whole.

These results have emerged from the same set of priorities that drove the Parkinson's disease work, namely the search for a robust source of cells that could be manufactured on a commercial scale. We've seen that one way to produce "scalable" cells is to start with progenitor cells from fetal brain tissue, and treat them with high concentrations of growth factors. Two other strategies emerged concurrently that have been successful in reaching the clinic. Both involve genetic manipulation. The first, "conditional immortalization," involves manipulating the cell cycle directly, and to understand this approach, we need to appreciate the process of cell division in a little detail.

The cell cycle describes the progression of a progenitor cell—or of any dividing cell—through the process of division. Before a cell can divide to give rise to two daughter cells, it must first replicate its DNA so that it can provide each daughter with a complete copy of its genome. Once this DNA synthesis phase—"S" phase—is complete, the cell can progress to actual division, the process by which the one cell becomes two. This is called "mitosis," hence this is the "M" phase. Once the cell has divided, the process can be repeated with a further "S" phase, and so on. Thus each individual cell generates its own little family tree.

In the gaps between successive "S" and "M" phases, the cell takes stock, and decides when the time is right to progress. Does it have the necessary energy reserves? Are all the cell constituents ready and in the right place? If so, the cell can throw the appropriate switches, and move on. One of the great achievements of cancer research over the last thirty years is that we now understand in considerable detail how these control mechanisms work, often involving genes called "oncogenes" that push and pull the cells around the cell cycle.

The idea arose that simply turning on an oncogene might be a way out the stem cell expansion conundrum. If the right oncogenes could be activated, then maybe the progenitor cells would divide without stopping. This strategy worked, and several groups reported that specific oncogenes, with colorful names like *c-Myc* ("see-mick") and *large T*, could indeed push fetal brain cells to divide without stopping. But a worry lay just below the surface:

once you had activated continuous division, how to stop it? Like Micky Mouse in Disney's *Sorcerer's Apprentice*, might you turn on the spell, but be unable to turn it off again?

The answer was to fit the oncogene with a switch. *c-Myc* acts to promote cell division by binding to certain genes in the cell nucleus and turning them on. Since these are genes that drive cells around the cell cycle, this has the effect of promoting cell division. But Gerard Evan and his team discovered that *c-Myc* itself could be fitted with a switch.[1] To activate genes *c-Myc* must get into the cell nucleus. To do this two c-myc proteins must combine into pairs, otherwise the c-myc remains in the cytoplasm doing nothing. The Evan team engineered a *c-Myc* variant that could combine into pairs only when it bound to a specific drug, "4-hydroxytamoxifen." This compound is related to the drug tamoxifen used to treat breast cancer, and Evan made use of an engineered piece of the estrogen receptor to build the switch. The consequence is that without the drug, the *c-Myc* stays inactive in the cytoplasm, but adding the drug moves the *c-Myc* to the nucleus, where it activates gene expression, and drives cell division.

John Sinden and colleagues at the UK company ReNeuron engineered this *c-Myc*–estrogen receptor construct into human fetal progenitor cells taken from cerebral cortex, and exposed the cells to 4-hydroxytamoxifen.[2] With the *c-Myc* switched on, the progenitor cells grew, and from the culture emerged clones of progenitor cells, each of which was expanded to give billions of cells. The researchers were then able to examine the properties of these emergent cell lines. In particular, they wanted to know that if they withdrew the 4-hydroxy-tamoxifen, the cells would cease division and differentiate. They found that many of these lines were indeed able to stop dividing and to start generating neurons and glia when the drug was withdrawn. So, by manipulating *c-Myc* expression, they'd been able not only to activate the two key stem cell properties—self-replication and multipotentiality—but also to put them under experimental control. One of these "conditionally immortalized" cell lines, called "CTX0E03," is now in clinical trials for stroke, as we'll discuss in a moment.

A second manipulation of gene expression used by researchers has been to manipulate the *Notch* signaling pathway. *Notch* is beautiful example of conservation in biology. It is a genetic pathway first discovered in fruit flies—the "notch" in question being the small indentation observed in the wing of flies with a mutation of the *Notch* gene—but *Notch* turns up everywhere in both vertebrate and invertebrate biology. What it and its associated genes do is regulate the social interaction of cells. In a population such as the progenitor cells we looked at in figure 3.1, the social behavior of the cells must be precisely regulated. All the progenitor cells are primed to make neurons, but they can't all differentiate simultaneously or the neuroepithelium would become depleted. Like a crowd exiting the stadium after a football game: everyone wants to leave, but they must do so in an orderly fashion; some first, others later. Similarly, all the cells are trying to leave the neuroepithelium and become neurons, but *Notch* provides a mechanism whereby as one progenitor cell starts to become a neuron, its neighbors are directed to make more progenitor cells instead and to sit tight and wait their turn. One can quickly see how activating *Notch*, therefore, could be a useful strategy for maintaining a progenitor cell population and preventing the cells from differentiating prematurely.

Toshiro Mimura and colleagues employed the *Notch* activation to generate stem cell populations for therapeutic use, though they rationalized the approach differently since they were pursuing "transdifferentiation." We'll discuss this topic later, but for now let's note that the cells they generated—"SB623 cells"—are also currently in the clinic to treat stroke. The progress and potential of these two therapeutic products—CTX0E03 and SB263—are what we need now to examine.

Therapies for Stroke

The old adage "Don't look a gift horse in the mouth" is particularly inappropriate advice for scientist. Scientists need to be skeptical, and with good reason: important findings rarely fall into scientists' laps, and if the unexpected does suddenly appear, it pays to give it a long, hard look.

How, then, to treat the finding that introduced this chapter? Neural stem cells are injected into the damaged brain of an experimental animal with the intention of replacing lost brain cells. After about four weeks, the animal shows functional improvement, and, after six weeks, by some measures, the animal's behavior is so improved it cannot be distinguished from that of an undamaged control. But when the animal's brain is examined histologically, not only are there few newly formed brain cells, the injected stem cells have almost entirely disappeared. There has been little cell replacement, and yet the animal has recovered.

I've implied already that this finding isn't quite believable. But the interesting fact is this: of the small number of cell therapies currently in trials to treat brain disorders, a substantial proportion are seeking to act in precisely this "hit-and-run" fashion, to invade the damaged brain tissue, elicit a therapeutic effect, then disappear.

When CTX0E03 cells are injected into rats that suffered a stroke several weeks previously, the animals begin to show improvements in sensorimotor performance. This can be seen, for example, using either the "foot fault" or the "sticky tape" test outlined in chapter 1. These functional improvements first become evident roughly four weeks after engraftment, plateau over the ensuing four weeks, then remain stable for the remainder of the study, typically twelve weeks in total. This finding is robust, and—as we've already noted—decidedly nontrivial. Not many therapeutic agents can bring about this degree of functional recovery in the damaged brain.

Armed with these preclinical data, ReNeuron sought authorization to commence a clinical trial using clinical-grade human CTX0E03 cells to treat ischemic stroke, and under the auspices of the UK Medicines and Healthcare products Regulatory Agency (MHRA), a phase 1 trial began in Glasgow in 2011.[3] As always, regulators focused on safety, particularly given that this was the first cell therapy product in Europe to reach clinical trial for a brain disorder. You might also imagine that the use of the oncogene switch caused some eyebrows to be raised. The question naturally arose: could researchers be sure the switch was not going to reactivate and turn the cells into a brain tumor? Some critics of the conditional immortalization approach have been quite explicit in this regard. In a 2015 review of brain cellular therapies, for example, Julius Steinbeck and Lorenz Studer concluded: "Given the safety concerns associated with immortalization and the availability of alternative strategies and cell sources, such a strategy should only be pursued with the utmost caution."[4] In fact, the CTX0E03 cells passed multiple toxicity and tumorigenicity tests both in culture and in experimental animals without raising any concerns. In fact, once the *c-Myc* has been turned off by the removal of the 4-hydroxytamoxifen, it has proved impossible to reactivate it, even with the drug. The reason is that, once deactivated, the gene undergoes "silencing," a phenomenon well known to molecular biologists, whereby the DNA becomes chemically altered (methylated) preventing reactivation.

The small phase 1 trial with eleven patients commenced in 2010 in Glasgow. All were male, over the age of 60. Each had suffered a stroke six months or more before the trial, leaving a lesion at least the size of a cherry. Each patient was left with residual, lasting disability. Using a stereotaxic device fixed to the head, surgeons were able to guide a needle directly alongside the damaged area and to make multiple injections of cells.

This phase 1 trial was designed to provide safety data and was by this measure a success. There were serious adverse events reported, but they were mostly as would be expected with an aged, unhealthy group of stroke victims. Other, more minor events, such as bleeding, were associated with the surgical procedure rather than with the injected cells themselves. The study was too small (and was never intended) to provide convincing evidence of efficacy. Moreover, it was open label (with no control group) and therefore subject to all the uncertainty generated by the placebo effect we discussed in chapter 4. Nonetheless, most patients showed some improvement by most objective measures, and most improvements were stable for the twenty-four months of the study. For example, some patients showed reduced spasticity or improved movement of limbs. On the Barthel index—a measure that scores activities associated with everyday living such as feeding, dressing, and climbing stairs—there was an increase across the patient group as a whole of 3 points on the 20 point scale.

These results were sufficiently encouraging for ReNeuron to apply for a phase 2 clinical trial. The interim results released in December 2016 again showed evidence of efficacy. The primary end-point of the trial was met, though with a slight slippage in the time scale. Fifteen of the twenty-one patients showed significant clinical improvement in at least one of efficacy measure. These results were sufficiently encouraging for ReNeuron to announce—even before this phase 2 trial was complete—that it would go forward to a phase 2b study. We currently await the outcome of that trial.[5]

The SB263 study, hosted by Stanford University and sponsored by SanBio, Inc., has followed a similar trajectory. Again, positive data from the preclinical studies led to a phase 1 safety study with eighteen patients. Again, the outcome was encouraging. No serious safety concerns were raised. The most serious adverse events were those associated with the surgery, such as bleeding or headache, and as with the ReNeuron trial, some patients showed improvements. In the most extreme case, one patient was able to get out of the wheelchair into which he had been confined by his stroke. But again, the numbers were small, the outcomes variable, and inconclusive. A larger phase 2 study is now under way.

So what's not to like? Well, recall that the entire rationale for the approach was that stem cells have the potential to replace the neurons and glia lost in the aftermath of the stroke. Yet little cell replacement takes place following the injection of either CTX0E03 or SB263 cells into experimental animals.[6] Whether the replacement that does occur is sufficient to mediate any positive functional effect is not known. Neither is it known whether there is any cell replacement in the human patients since (thankfully) none of them has died yet. So what's happening, and should we be concerned?

In fact, the initial cell replacement rationale for both studies was based on a wild extrapolation and a misconception.

The extrapolation emerges from the extent of the challenge. The strokes suffered by the patients in each trial were typically between a marble and a golf ball in size. In the preclinical studies, the hole left by the stroke in the animals' brains was much smaller, but proportionally far greater. Indeed, a stroked rat might lose up to 60 percent of one hemisphere—certainly enough to kill a human. Billions of brain cells were lost in the rats, and yet they showed recovery following the injection of less than half a million cells. Patients, being larger, got more cells—up to ten million in the current studies—but in each case, even if all the injected cells had differentiated into neurons in the host brain, probably less than a tenth of one percent of all cells would have been replaced. Hard to imagine that this would have been sufficient to bring about functional recovery. Indeed, this would be roughly equivalent to the degree of natural cell replacement observed in both the 2000 Macklis, Magavi, and Leavitt study and the 2002 Lindvall study that we discussed in chapter 4—replacement we've already judged to be insubstantial. It follows that even if the cell replacement had been 100 percent successful, it would probably have been inconsequential.

The misconception relates to a phenomenon called "transdifferentiation." We've seen that the CTX0E03 cells are conditionally immortalized neural stem cells that have a genuine capacity to generate neurons and glia, whereas SB263 cells are not neural at all. They are derived from cells that go by multiple names—mesenchymal stromal cells, mesenchymal stem cells, bone marrow stromal cells (all usually referred to as MSCs)—and are derived from multiple sources (bone marrow, fat tissue, or umbilical cord blood). They became linked with stroke because they were thought to be capable of transdifferentiation into neural cells.

If MSCs are injected into the damaged brain, then a small number of them do appear to differentiate into neural cells. This was quite a stunning observation when it was first reported because embryologists at the time were quite certain that this was impossible. One of the earliest "fate decisions" made by cells during development is which "germ layer" they are going to belong to. As we saw in chapter 6, neural tissue is derived from the ectoderm, the skin germ layer. The stromal cells are derived from mesoderm, the layer that gives muscle, blood, and most connective tissue. While it seemed inconceivable that they could somehow swap over to become neuroectodermal, this was in fact what several researchers reported. These mesodermal cells, they believed, were turning into neurons and glia.

Whether any such transdifferentiation does actually take place has never been entirely resolved though it is reasonably clear that much of what was observed was actually artefactual. Some of the injected stromal cells were actually fusing with host brain cells, thereby giving the appearance of transdifferentiation. In any case, trivial numbers of new neurons were being generated, which brings us back to the extrapolation problem. Could so few new neurons really be having a significant effect?

Nonetheless, there was an opportunity here. Whereas neural stem cells were hard to come by, the stromal cells could be relatively easily isolated and expanded. The *Notch* manipulation we noted earlier was first adopted because it was thought to encourage the transdifferentiation of MSCs into neurons, rather than expand the MSC population, though this is probably its real virtue. As a consequence, these *Notch*-manipulated MSCs found their way from preclinical studies into clinical trials.

None of this helps us explain why the animals show functional improvement, or why we have any hope of clinical benefit for patients with this approach. If they are not replacing neurons, what are these cells actually doing?

Reinterpreting Stem Cells

The key realization was that scientists had adopted too narrow a view of stem cell function. Stem cells had been defined as being self-replicating and multipotential, but in fact these are two aspects of a broader unifying property: stem cells are responsive. This was implicit in our discussion of hematopoietic stem cells in chapter 2: they sit in their bone marrow niche, judge what is required of them, and act accordingly. What researchers had missed is that the stem cells can do more than just make more cells. They can react in other ways, modulating their response to a changing environment.

My colleague Michel Modo did an interesting experiment. He used a brain imaging technology called "pharmacological MRI" to look at dopamine signaling in the striatum.[7] Dopamine is one of the principal neurotransmitters that regulate function in the striatum. If the tissue was damaged by a noxious stimulus, cells died over time and consequently there was a loss of the dopamine function in the striatum. But if neural stem cells were injected into the striatum, this loss of function was dramatically reduced. There were no new neurons formed by the injected cells, but the integrity of the tissue was somehow protected, what neuroscientists call a "neurotrophic effect." This gives us an inkling of what the stem cells are actually doing. Somehow the injected cells are protecting the injured brain from ongoing damage. So how is the injection of the stem cells achieving this effect?

We saw in chapter 2 that the reaction of the brain to damage is complex involving not only the brain tissue itself, but also the immune, circulatory, and endocrine systems. It turns out that injecting stem cells into this milieu modulates these responses. The stem cells secrete factors that act on the cells around them, and help orchestrate a more effective damage response. Simply, brain appears to react to injury more effectively if it is injected with stem cells. The cells secrete factors that change the behavior of the surrounding tissue, a process biologists call a "paracrine effect." Clearly, this effect is pretty powerful if it can orchestrate brain repair; and it is this observation—that the brain can be induced to more effective repair—that is the real lesson to have been learned from these studies.

So what is actually changing in the brain? In fact, so much changes in response to stem cell injection that the challenge is to disentangle the multiple effects. The most readily grasped effect is the formation of new blood vessels. Several stem cell types secrete factors—vascular endothelial growth factor (VGF), angiopoietin, and others—that promote the formation of new blood vessels.[8] Intuitively, this seems likely to aid brain repair. Since a stroke is caused by the interruption of normal blood flow, anything that improves blood flow seems likely to aid recovery, although how much impact this might have many months after the stroke is yet to be clarified.

Less easily understood are some of the other effects that have been observed. Engraftment with stem cells causes the host neural stem cells to generate more new neurons. We've dwelled more than once on the observation that the generation of new neurons in the stroked brain is inadequate. Enhancing this inadequate response again seems intuitively to be a positive change, but how it might actually help is similarly unclear. In the 2012 study my colleagues and I conducted, grafting "CTX0E03" cells into a stroked rat led to a roughly sevenfold increase in new host-derived neurons surviving in the striatum four weeks after the graft.[9] Sevenfold is a big change, but if the normal situation—replacing less than 0.01 percent of the neurons—doesn't help, would 0.07 percent make any difference?

More promising is immunomodulation. Microglia, the resident macrophages of the brain, converge on damaged brain tissue, combating the destructive forces unleashed by damaging agents such as ischemia. The role of microglia in neurodegeneration has come under intense scrutiny in recent years, and these cells are now considered to play a more nuanced role than had been previously thought. The properties of the microglia themselves change with time, and their influence matures as the acute response to injury modulates from damage limitation to repair. This can be seen most readily in the factors that the microglia themselves secrete and the action of these factors on the damaged brain. Immediately after damage, the microglia encourage inflammation—the acute protective response observed in all injured tissues. Later, however, they switch to the production of factors such as vascular endothelial growth factor (VEGF), insulin-like growth factor 1 (IGF1), and brain-derived neurotrophic factor (BDNF), all of which encourage neurogenesis, synapse formation, and tissue remodeling. Several studies now suggest that there is a positive interaction between engrafted stem cells and these microglia, and this may be what drives the neurotrophic effect. There are more active microglia in brains injected with stem cells. This mutual reinforcement sounds more like the amplifying regenerative mechanism that might support a therapeutic effect, so while the evidence is still only circumstantial, it currently looks like our best explanation.

My guess is that there is much more to be discovered regarding the interaction of stem cells with damaged tissue, and the mode of action of the cells may well turn out to be immensely complex. But while the details are still unclear, the hit-and-run effect seems to be this: that though the engrafted cells only survive a short time, they are able to sense how best to react to the damage, and then push the tissue into a more effective mode of response. Just like a good coach with a sports team can make tactical changes at half time, almost imperceptible to spectators, that shift a team stuck in defense into attack, so the engrafted cells can change the balance of tissue physiology. And once the balance is shifted, the effect can be sustained. The team keeps attacking even if the coach isn't on the sidelines for the second half of the game.

How might this unanticipated therapeutic opportunity play out? Much hangs on the next round of stroke trials from SanBio and ReNeuron. Though their structure is not yet known, later-phase trials will certainly include a placebo arm. We will then discover whether this hit-and-run approach actually works. Equally important, we will learn how much it works. Is it sufficiently efficacious to justify the cost, commitment, and risk associated with such an invasive surgical procedure? This is, after all, precisely the point at which the Parkinson's fetal cell studies crashed. Can this generation of cellular therapies survive the comparison with placebo?

If the phase 3 clinical trials succeed and regulatory approval follows, these would become the first licensed cellular therapies for a brain disorder worldwide. Though this would not be the end of regulatory concerns, the significance of this milestone would be hard to exaggerate. We'd discover how the therapies perform in the real world, which patient groups benefit, and whether there are any unforeseen outcomes. What's more, the range of disorders to be treated would start to stretch. While the therapies will only be licensed for the indication for which they were tested—disability resulting from stroke—clinicians would want to discover whether other patient groups might benefit. Hemorrhagic stroke and traumatic brain injury would be the obvious next stops.

Adoption of the new therapies will also depend on some critical nonclinical issues, particularly on some complex pharmacoeconomic issues. The first therapies will be impossibly expensive, with small scale and high cost-of-goods. Bringing down the costs will be crucial in order to match what health-care providers are prepared to reimburse, and to get approval from bodies such as the UK National Institute for Health and Clinical Excellence (NICE). The economic barriers will at first appear insuperable, but like vaccines and monoclonal antibodies before them, cellular therapies (if they work) will surely be made cheaper and more accessible. Nonetheless, these non-clinical barriers are not trivial, and the success or otherwise of cellular therapies will be determined as much by non-clinical issues as by the strength of the science.

A second outcome is that these therapies will immediately become the benchmark against which other therapies will be measured. It is conventional in clinical trials to compare the novel agent not only with placebo, but also with an appropriate comparator, usually the conventional treatment for that particular disorder. That's not happening in these stroke trials because there is no licensed medicine for stroke disability. This unmet medical need is precisely why these novel therapies are so attractive. But, once licensed as medicines, the therapies would become the comparators for future studies.

Most of the rest of this book will address the question: Where do we go from here? Specifically, are there prospects for real cell replacement? Hit-and-run is fine if it works, but the question still remains: could we actually replace lost brain cells and thereby bring about a substantial improvement in function? But note: if the current trials are successful, cell replacement would not only have to do better than placebo, it would have to do better than hit-and-run.

What's more, the discovery of this hit-and-run neurotrophic mechanism has has set other hares running. If stem cells can bring about this effect, couldn't other agents do the same? Probably only cells are capable of this clever response, whereby they monitor tissues and manipulate the outcome. Nonetheless, perhaps the factors that the cells produce might be druggable. Could these factors themselves become medicines, obviating the need for cell engraftment?

In fact, the factors the cells secrete have been known for a long time. Several are already in use therapeutically particularly in blood disorders (targeting our now-familiar hematopoietic stem cells), but also for tissue damage, for instance skin, muscle, and joints. They have also been trialed in the nervous system, but there are a couple of problems here. First, the brain is protected by the "blood-brain-barrier." This high-level filtration system prevents many blood components having access to brain tissue. Proteins (such as growth factors) tend to be kept out, which makes them difficult to administer. Second, growth factors are "pleiotropic": they do different jobs in different parts of the body. So, if they are introduced into the blood stream like a conventional drug, they often struggle to get into the brain, and instead interfere with some other process elsewhere.

The stem cells themselves offer one clever solution to this drug delivery problem. The paracrine mechanism underpinning the hit-and-run effect is probably mediated not only by factors the stem cells secrete but also by structures called "exosomes" that they also secrete. These are tiny pinched off blobs of cell. They disperse and fuse with other cells, transferring material from their cell of origin to neighboring cells. Since exosomes are, in essence, little sacks of cytoplasm surrounded by cell membrane, they contain cellular components. Some components become particularly enriched in the exosomes as they are produced. These include some RNAs, both protein-encoding messenger and regulatory RNAs. So, effectively, exosomes are tiny control centers that can be secreted by engrafted stem cells and taken up by neighboring brain cells in the process of recovering from injury.

It appears that exosomes might be important vehicles mediating the paracrine effect, which raises the possibility that they themselves might be useful therapeutic products. They could be collected from stem cells in the laboratory, concentrated and purified, then delivered to the patient. They could be engineered to amplify and modify the payload they carry, making them enormously promising potential therapies of the future.

In this chapter, we have met the most advanced cell therapies for stroke. Probably the first such medicines to gain regulatory approval will emerge from this pipeline. These are not, however, the only efforts ongoing. In the next chapter, we shall widen our scope to look at other initiatives in other disorders.

8 Feral Therapies

The scenario that has emerged over the last few chapters is an odd mix of success and failure: the advancing hit-and-run approach, but the stuttering cell replacement strategy. The story of the rest of the book is primarily about where the replacement strategy goes next, but before we look at what further tools we can find in the toolbox, we need to take stock.

A strong case can be made that current stem cell therapies for brain disorders are doing more harm than good. Francis Collins, director of the US National Institutes of Health, is one of many to draw attention to some pretty unsavory practices. There were the three women in Florida with age-related macular degeneration who each paid $5,000 to be injected in both eyes with stem cells taken from their own fat tissue, only to end up with severe loss of vision.[1] Or the Australian woman who died following injection with stem cells sucked from her own abdomen to treat her dementia.[2] Or the 66-year-old man who suffered back pain, paraplegia, and incontinence as a consequence of the tumor on his spine that emerged following stem cell infusion into nervous system.[3] All of these cases are tragic for those involved, but surely the most absurd is the 57-year-old man in the United States who ended up in intensive care with paralysis and blindness following an injection of stem cells into his scalp...to cure his baldness.[4] Any informed scientist or clinician looking at these cases would conclude that their outcomes were not the result of bad luck. They were ill-considered and should not have occurred.

Some of these are examples of "stem cell tourism," patients seeking experimental therapies in countries with inadequate regulatory oversight. Finding a clinic willing to provide such quack therapies is not difficult in the internet age. Clinics such as the Wu Center in Beijing will treat your motor neuron disease, diabetes, spinal cord injury, eye disorder, epilepsy, and much beyond with uncharacterized, nonstandardized stem cell extracts.[5] Perhaps more shocking, patients don't have to go that far to find "direct-to-consumer" marketing of unregulated stem cell therapies. Companies have cropped up in Australia, the Netherlands, and Germany, though many have subsequently been shut down by the authorities. A report in 2016 identified 351 US companies marketing stem cell therapies directly to consumers, many noncompliant with federal regulations.[6] For most of these therapies, there would be little evidence that they work. Certainly, they would not have undergone appropriately constructed clinical trials. Worse for the authorities, it transpires that the ClinicalTrials.gov website, introduced by the US National Institutes of Health (NIH) to provide transparency and accountability to clinical trials, is being abused as a marketing vehicle for unlicensed cell therapies, with companies taking advantage of the failure of the NIH to screen database entries, thereby providing a veneer of respectability for companies who are charging for participation in supposed trials of unregulated therapies.[7]

How has this situation arisen? There are many contributing factors, not least what the International Society for Cellular Therapy (ISCT) has called the "near magical hold" that stem cell therapies have in the eyes of patients. But in scientific and regulatory terms, it is primarily the consequence of the odd trajectory that scientific progress has taken. The primary objective of stem cell therapy—cell replacement—has not been met, an embarrassment in this area of promissory science, which like gene therapy before it has been conspicuously overhyped. The unanticipated hit-and-run strategy looks promising and is being pursued by both commercial and academic advocates—correctly in my opinion since it offers a genuine prospect of success. But our lack of understanding of the underlying biology, particularly our poor grasp of the mode of action, has opened a Pandora's box. No one quite knows where the boundaries of this opportunity lie. This oddly ambiguous outcome has created a vacuum, and what has swept into that space is scientifically and ethically dubious. Having taken this tangential direction, the technology has left regulators uncertain of which levers to pull. Too many players—governments, learned societies, practitioners—have been compromised by their desire not to be left behind. The outcome has been a collective failure to protect patients.

The Science Gap

To understand how this sharp practice has arisen, we need to recognize where the gaps in our knowledge lie. First, let's note that in one sense this book has entirely concentrated on the tip of an iceberg. Our subject is brain repair as delivered by cell therapies, but cell therapies extend much more widely than the brain. In 2016, there were 804 clinical trials of regenerative medicine products worldwide according to the Alliance for Regenerative Medicine.[8] Of these, only 54 (fewer than 7 percent) were for brain disorders. Almost half were for cancers, an area where cell therapies are beginning to make a genuine impact.

Perhaps the most eye-catching number is the 556 clinical trials of mesenchymal stem cells (MSCs),[9] now involving more than 2,000 patients.[10] MSCs are the cells we met in the context of the SanBio trial in chapter 7. What is remarkable is that these cells are also in trials for disorders as diverse as graft-versus-host disease, type 1 diabetes, and autism. We've seen how this approach can have powerful effects in preclinical studies of stroke, and the first hint of a clinical effect in patients. But is it spreading too far too fast—really, baldness?

If we look closer, we find three fundamental scientific problems. First, MSCs are very poorly defined. They were originally characterized as bone marrow cells with the capacity to generate bone, cartilage, and adipose cells. Similar cells have since been identified in a great diversity of tissues—dental pulp, amniotic fluid, and dermal tissue, to list just three—but science cannot currently say just how similar these different populations are. The International Society for Cellular Therapy (ISCT) attempted to clarify this situation by defining MSCs as cells possessing certain precise molecular and developmental characteristics.[8] While this set a standard, cells so defined are still a mix of different phenotypes. Moreover, the ISCT classifies MSCs as derived exclusively from bone marrow, but nothing prevents purveyors of MSCs from defining them as broadly as they wish. Thus the problem has been moved forward but not solved.

This leads us to problem number two: the mode of action of MSCs is obscure. Although originally defined by their developmental potential, MSCs are almost certainly working (where they're working at all) in a hit-and-run fashion. Several mechanisms of action have been proposed, including immunomodulation and tissue remodeling, and their impact is undoubtedly broad. They're known to secrete a range of factors, but in almost no therapeutic situation, even in animal model studies, is the active factor known for certain. So, the most obvious question a researcher would be asked—are your cells making the therapeutically active substance?—cannot be adequately addressed? This is why the anarchic, hit-and-run approach is so unsatisfactory. In none of the cases cited above would the active factor have been known, and therefore it could not have been assessed—even had there been a willingness to do so—prior to the injection of the cells into each patient.

The third problem is that there is no standard way to handle MSCs. One reason why they've become so popular is that they are relatively easy to grow. We've already noted that other stem cells can be difficult to expand in culture. Not so MSCs. What is also true, however, is that even relatively small variations in culture conditions are likely to lead to substantially different outcomes. Again, this fundamental problem is simply ignored by those currently promoting MSC therapies.

Patients enrolling for current MSC therapies are playing a particularly crazy form of Russian roulette. They are taking a shot of stem cells. Most chambers in the metaphoric revolver are almost certainly empty; that is, there are no efficacious cells in the mix, or at least not enough to do any good. There might be a chamber that actually delivers something of value (Feeling lucky, Punk?), but more likely is a chamber containing a bullet that will blow the top of your head off.

These problems together put regulators in an enormously difficult predicament. They know that the production of cells for clinical trials is not as tightly controlled as they would like, but they don't want to kill this field of endeavor at the outset by being too restrictive. This predicament comes into clear focus with the issue of potency assays. Regulators—such as the Food and Drug Administration in the United States or the European Medicines Agency in the European Union—would like to see a clear measure of potency applied to any therapeutic product. They want to be sure that each batch of cells destined for patients has equivalent effect. Moreover, they want the potency of the batch of cells going into patients to be equivalent to the batch employed for the preclinical research, otherwise there would be no correspondence between the preclinical and the clinical studies. Such a requirement presumes that the mode of action is understood. Regulators would not necessarily expect the precise mode of action to be understood in full at the outset, but they'd want to see a program of research that would lead to such an understanding during product development. Consequently, the initial potency assay could be relative permissive—pick the factor you think most likely to be important and measure it. But before trials are complete and the therapy is licensed, regulators would want to see a defined production process with assays to match.

No cell therapy products have yet met this gold standard. Most products in early-stage clinical trials are some way short of this goal, and many may fail as therapeutics, not because the cells don't work, but because they can't be brought up to regulatory standards. This was part of the problem with the StemCells, Inc., failure we discussed in chapter 6. There was a discrepancy between cell batches, and it was unclear whether the cells that went into patients were equivalent to the cells that had worked in the preclinical studies.[11] As developers in this field are fond of saying: "The process is the product."

Beyond Regulation

All of this might be manageable were regulators judiciously empowered, researchers appropriately critical, and clinicians sufficiently conscientious. There are two particular issues—neither unique to cellular therapies—that together have created a spectacular regulatory hole. The first is the "hospital exemption" afforded by most jurisdictions to doctors who wish to treat an individual patient with an innovative therapy. A doctor can administer an unproven medicine to a seriously ill patient provided no existing medicine has been licensed for the relevant indication and, in the doctor's judgment, there is a reasonable probability of a positive outcome. This allows a shortcut through the registration process. Whereas a pharmaceutical company must subject a novel therapy to years of testing and clinical trials before it can be registered, the same therapy can be offered with no prior testing at all provided if a clinician deems to appropriate. Certainly, an internal approval process would operate in the hospital in which the doctor operated, but in many countries beyond this, there is little external control.

There are multiple constraints on the application of hospital exemptions: the use of the therapy must be nonroutine, individualized, and employed within a hospital setting. But even within an extensively regulated environment such as the European Union, considerable variation exists between different member countries.[12] And in many countries, clinicians may have considerable latitude on how to apply the exemption. If they decide to offer to graft some stem cell therapeutic into a patient's brain for $10,000, then so be it.

The second issue is provision of autologous cells. Most of the therapies we've considered so far have been "allogeneic," meaning that the donor of the cells is not the same individual as the recipient. For example, the fetal cells for the Parkinson's disease treatments we discussed in chapter 5 came from an aborted human fetus and were grafted into a Parkinson's patient. The alternative approach is termed "autologous," where donor and recipient are the same. This has a much longer history in stem cell medicine and is currently in routine use, particularly for blood disorders. We discussed such an example in chapter 2. A child with leukemia has bone marrow stem cells collected prior to radiation therapy. Then after the radiation, the cells are reinfused into the patient, where they repopulate the blood with noncancerous cells.

Like therapies permitted under the hospital exemption, autologous cell therapies are regulated more leniently than their allogeneic counterparts. It would make no sense to require each patient's cells to go through the entire regulatory procedure before the child could be treated. So such therapies are not considered to be advanced therapy medicinal products and these autologous therapies do not require such regulatory approval.

But what if bone marrow stem cells are being injected, not back into the blood system, but into the brain? What if—as in the Florida example that began this chapter—stem cells are taken from a woman's fat tissue, then injected into her eyes? Neither US nor European regulators would consider this autologous because as well as being the patient's own cells, an autologous therapy should be "minimally manipulated" while out of the body, and when reintroduced should perform essentially the same functional role as before it was extracted—or in FDA-speak: their use should be "homologous." Thus, if the extracted cells were blood stem cells, then their use is only autologous if, following engraftment, they reoccupy the blood-forming niche in the bone marrow.

A Seminal Case

There is, however, little consensus currently on how these definitions and regulations should be applied. No case better exemplifies the lack of resolution than the US District Court decision in 2012 against the Colorado-based clinic Regenerative Sciences.

In 2012, Regenerative Sciences was offering a therapeutic product called "Regenexx-C"—essentially, an MSC product—for the treatment of arthritis and a number of other orthopedic conditions. Bone marrow stem cells were taken from patients, expanded in the company's laboratories, treated with antibiotics, combined with excipients, then reinjected into the site of orthopedic damage in the original cell donor. Regenerative Sciences had treated over 800 patients with this product and claimed its Regenexx-C cells were safe. But the company had no systematic data to show efficacy, nor had it conducted a clinical trial.

In 2008, the Food and Drug Administration gave notice that use of Regenexx-C might be in violation of federal regulations and laws, in effect, defining the product as a drug and not an autologous therapeutic product. Disputing this definition, Regenerative Sciences insisted that it was simply engaging in the "practice of medicine" and that its Regenexx-C therapy fell outside the FDA's remit. After a number of lawsuits were filed and counterfiled, a US District Court finally sided with the administration in 2014, and this supposedly autologous cell therapeutic was officially deemed a drug.[13]

This has been an enormously controversial judgment and lies squarely in the difficult terrain we have been surveying. Are the regulators holding back innovative medical practice or protecting vulnerable patients from untested therapies? Even before that judgment was handed down, many commentators had jumped in to claim the former. "The FDA's Misguided Regulation of Stem-Cell Procedures: How Administrative Overreach Blocks Medical Innovation" read the title of a 2013 report by the Manhattan Institute.[14] And a 2011 *Wall Street Journal* piece decried the FDA's "impulse to regulate, and thus forestall, cell therapies used to help repair damaged body parts."[15] For these and other like-minded commentators, the FDA regulators were interfering with progress and needed to back off.

Much of the argument over the judgement centered, first, on whether the FDA had jurisdiction over this aspect of medical practice and whether it had behaved appropriately. There were accusations that the FDA had changed its definition of "autologous" without due process, and also that the Regenexx-C activity involved no "interstate commerce" and therefore fell outside the agency's federal purview. These issues are of legal interest, but they don't address the core question of whether the FDA action was in the interests of patients. The biomedical component of the judgment hung precisely on whether the Regenexx-C cells were minimally manipulated and serving an homologous function. In regard to the former, the FDA asserted that, since the cells had been mixed with blood products, expanded using a variety of reagents, and handled extensively in a laboratory setting, then, no, they were not minimally manipulated, and that Regenerative Sciences had in fact manufactured a product. That the company had its own (small-scale) production facility for this process seemed to support this assertion. For many, however, this was an extreme interpretation of the "minimal manipulation" rule. If cells could not be expanded even for a short period in culture, then minimal manipulation really meant minimal. Conversely, it didn't help the company's case that the FDA had inspected its facility on two occasions, and found it wanting. Steps had not been taken to assure that the product conformed to "appropriate standards of identity, strength, quality, and purity" Not only was Regenexx-C a manufactured product; it was a poorly manufactured product. Moreover, Regenexx-C failed the "homologous test." Bone marrow stem cells produce blood. They don't repair damaged joints, at least not as far as scientists know.

So was this an example of the FDA's misgivings about "anything novel in medicine," or was the administration fulfilling its duty to protect US citizens from unproven, and potentially dangerous treatment? Though the barbs aimed at the FDA were many and various, two stand out as pivotal. For one, the FDA's approach was stifling biomedical innovation. For another, the FDA was simply not doing what it was supposed to. Or, as Mary Ann Chirba and Stephanie Garfield of Boston College Law School tellingly put it: "The FDA should recognize that it makes little sense to impose a regulatory framework developed for mass manufacturers on small physician practices."[16]

The first of these objections would carry more weight were it substantiatiated. The fact that bioscience was being driven offshore is actually poor evidence though frequently cited. Much of this activity is certainly taking place where regulations are looser. The core question is whether unlicensed bioscience applications are truly advancing the field or merely advancing profits at the expense of vulnerable patients. As we've seen, Regenerative Sciences had treated 800 patients. This represents neither an investigative sample nor the treatment of a small number of desperate patients, as envisaged under the "hospital exemption," but, rather, a full-blown commercial enterprise. The company had not attempted a proper clinical trial, without which the supposed efficacy of its Regenexx-C treatment would be forever uncertain. And critics might be forgiven for thinking that a slight conflict of interest was holding it back from such trials. The company was reportedly charging up to $54,000 per patient for the unproven therapy.[17] Why risk a clinical trial?

In fact, what's missing in this field isn't innovation but careful application. We saw, in chapter 5, that the Parkinson's treatment trials entailed a multitude of variables, each of which had to be precisely honed. Researchers don't generally get lucky in science: they have to work things out properly. If they need to guess how many cells to inject, or where to put them, or how to process them before they inject, then, more often than not, they'll guess wrong. The truly innovative enterprises are those we've encountered in earlier chapters that are seeking to establish their novel therapies on firm scientific foundations.

As for the second objection: should the FDA demand the same high standards from a small clinician-led enterprise that it does from Big Pharma? Well, if you were a patient being treated at the Regenerative Sciences clinic, wouldn't you want to know that the cells you were receiving were pure, that the clinicians had made sure that your cells hadn't been mixed up with anyone else's, that the reagents they'd used were free from contaminants, and that there was nothing in the injection that shouldn't be there? Wouldn't you want the same standards to apply that apply to all approved medicines? Many of the companies we have met in this book are small, yet they seek to meet these regulations. Why should Regenerative Sciences be exempt?

It is true that the codes need to be applied sensitively, and certainly ham-fisted regulation can needlessly encumber small businesses. Chirba and Garfield make an interesting comparison with the support given within the European Union for small and medium enterprises (SMEs) to ensure they can negotiate the early development phase of novel drug discovery.[18] This is a view I would strongly endorse. Nonetheless, this is actually a different point from whether regulation is appropriate. Certainly, small businesses need help, but no one benefits except their shareholders if they're given a free pass.

The Regenerative Sciences case is an interesting example precisely because it is not extreme. The company wasn't blinding its patients with fat cells. Nor was it killing its patients with brain tumors. As far as I can judge, it was a genuine enterprise trying its best to improve the lives of its patients. The question is only whether its patients—all patients—would've been better served had the company adopted a rigorous approach both to cell manufacture and to clinical practice. Shouldn't we expect all providers of therapies to aspire to those standards?

Trials and Tribulations

The case for a systematic approach to clinical trials for advanced therapies would be easier to make if they were not so fraught with difficulties. Many of the problems revolve around the concept of "clinical equipoise."

We've seen clearly the problem with employing therapies that have not been properly trialed. You simply don't know if the patients are doing better than they would have done without the treatment. This uncertainty is exacerbated with stem cell therapies, where the patients' belief that stem cells are "near magical" heightens the potential for a strong placebo effect. We've seen that most therapy trials commence as phase 1 safety studies followed by "open-label" phase 2 trials, where there's no control group, and where efficacy is measured against what might have been expected from this patient cohort based on previous clinical experience. There is broad (though not universal) agreement, however, that efficacy will eventually have to be tested in placebo-controlled trials, where one group of patients receives the therapy, and a control group has "sham" surgery, that is, they are treated identically to the experimental group, receiving anesthetic, having a hole drilled into their heads, but receive no injection of cells. This is the ethically problematic core of controlled trials. How can this treatment of patients be justified?

One answer lies in "informed consent." After the structure of the controlled trial is explained, patients who sign up do so in the knowledge that they may—or may not—get the active substance. They also know that they may not benefit from the treatment themselves, but that the trial and their participation in it are necessary to validate the therapy for the future benefit of others. Thus altruism lies at the heart of the controlled clinical trial.

But this still isn't good enough. Clinicians have to justify their own role in the process: How can they allow patients entrusted to their care to undergo a placebo procedure that they don't believe will help while denying them an active substance procedure that may?

Bioethicists have employed various formulas to address this question, but most settle on the concept of "clinical equipoise." Administering a placebo is deemed ethical if there is "genuine uncertainty" whether the active treatment—the stem cells in this case—will produce a more beneficial outcome for patients than the placebo. Most ethicists would also insist that a controlled trial should be so structured that, by the end of the trial, the clinicians could judge whether the active treatment was in fact more beneficial. Otherwise, the risk taken by patients would be for naught.

The first problem is, who's to judge whether there's equipoise? Presumably, the clinicians at Regenerative Sciences thought that they did know that the cells were better than nothing, and advised their patients accordingly. Dr Wu in Beijing presumably believes that he can cure anybody who walks through the door, whatever their disorder. If he doesn't believe there is equipoise, if he is confident the cells are better than nothing, then surely it would be unethical for him to conduct a trial thereby denying some of his patients access to this wondrous therapy.

Bioethicists such as Charles Weijer and colleagues at Dalmousie University have argued that equipoise has to be a community decision. "Clinical equipoise…recognizes explicitly that it is not the individual physician but the community of physicians that establishes standards of practice."[19] But this assumes consensus, and that's clearly what's missing. What we can readily imagine, however, is that patients' grasp of equipoise will not concur with that of doctors. Patients may or may not believe that stem cells are "near magical," but if they join a trial, they want the cells, not the placebo. A study of Parkinson's patients and their relatives by Teresa Swift showed that, although most patients agree that a control group is necessary, they don't want to be in that group.[20] They would only feel inclined to sign up for the trial if they were in the active group, receiving the stem cells. They are not "in equipoise." Moreover, patients move further from equipoise the more severe their condition. Patients join a trial precisely because it is their only access to a stem cell therapy.

This, of course, is the key to the success of stem cell tourism. The sicker the patient, the more desperate: the more desperate, the less altruistic. For terminal patients, the cells are the last chance, and balanced judgment goes out the window. How meaningful, then, is "informed consent"? Organizations such as the International Society for Stem Cell Research (ISSCR) are trying gallantly to assist patients to reach informed decisions before pursuing stem cell therapies. The Society's website features its *Patient Handbook on Stem Cell Therapies*, available in ten languages, as well as a number of other guidance tools.[21] Efforts in this direction have not been overly successful as yet, but in the absence of effective regulation, education is the only tool in our armory. Just as "There are no atheists in a foxhole," so I suspect there are few stem cell nonbelievers among the terminally ill.

It might be argued that I've concentrated on the failures of unregulated autologous therapies without acknowledging their successes. The websites of the direct-to-consumer providers are replete with testimonials from satisfied customers who thank the clinics for saving their lives. There's no question that a subset of patients come away grateful for the therapy they've received.

We have already touched on the difficulty of interpreting such data, where there's a confluence of three factors: uncertain diagnosis, incomplete follow-up, and poor numbers. Not knowing what proportion of patients from any particular clinic with a particular diagnosis show improvement, we can't compare that number with what would be expected from the natural progression of the disorder. For example, many patients with multiple sclerosis, do experience remission, only to regress again later.

In properly structured clinical trials, the assessment criteria are specified in advance. Clinicians have a range of tests to choose from, many of which are equally valid. Nonetheless, they need to specify in advance which test they'll apply and stick to it. Why? Because if they do enough different tests on enough patients, some will eventually come out positive, regardless of whether there was any genuine benefit from the treatment. Clinics using unlicensed stem cell therapies rarely stick to this discipline. They broadcast a positive outcome however many negative outcomes accompanied it.

Without proper clinical management and follow-up, testimonials are of no value in estimating the actual efficacy of a novel treatment. Even disorders that normally have an unremitting progression throw up anomalies. Stephen Hawking, diagnosed with motor neuron disease in 1963, was told he'd probably survive no more than two years, that being the norm for those afflicted with this disorder. He died in 2018, more than a half century later, at the age of 76, having thankfully and inexplicably defied all expectations. Had he received a stem cell therapy as a young man, no doubt we (and he) would have acclaimed his recovery as a medical breakthrough.

A big factor surely is cognitive dissonance. Friends and relatives raise thousands of dollars to send you to Beijing to receive this wonder drug. They see you off at the airport amid hugs and tears, with a local TV crew on hand. Are you seriously going to come back saying you feel the whole thing was a waste of time and money?

I followed up informally a case in the United Kingdom, where a patient had traveled to China to receive stem cells for treatment of motor neuron disease, and subsequently posted a glowing testimonial on the clinic's website. A relative contacted subsequently confirmed that the patient had in fact died only a few months after returning home, as would be expected through the normal progression of the disease. Apparently, however, the National Health Service was to blame for the patient's death, according to the relative, because of the poor care received on return. The patient had remained convinced of the success of Dr. Wu's treatment right up to the end, and the testimonial remains on the clinic's website, an apparent "success," with no further follow-up.

Naturally, extrapolation from my little failure narrative is no more valid than from any individual patient's success story, and I am, of course, completely outnumbered. But those patients who have been followed up systematically have universally failed to show the improvement claimed by themselves and their stem cell therapists. For example, Bruce Dobkin, Armin Curt, and James

Guest followed up on seven patients who had traveled to Beijing for stem cell therapy for spinal cord injury. Of these, they observed, complications, including meningitis, occurred in five patients, and "No clinically useful...improvements were found."[22] It is difficult to argue when they conclude with the ISSCR that "treatments are sometimes exaggerated by the media and other parties" and "by 'clinics' looking to capitalize on the hype by selling treatments to chronically ill or seriously injured patients."[23]

Internationally, scientists and clinicians are calling for stricter, enforceable rules.[24] How can we regulate this market and protect vulnerable patients without holding back progress, while acknowledging that we don't understand the science well enough to derive appropriate standards and controls? I would highlight several factors. First, regulators can apply appropriate pressure through their licensing of facilities. Of all the observations made by commentators in this field, the one I disagree with most is the contention of Chirba and Garfield that the highest standards should not apply to small-scale production. The most damning indictment of Regenerative Sciences was that, when inspected by the FDA, its facilities were not up to standard. Whatever the limitations of our knowledge, we do know how to make cell products that are clean, pure, and well manufactured. There's evidence, as I write, that the FDA is scaling up its surveillance, with the release of new guidance documents for regenerative medicine,[25] acknowledging that its regulators have "seen products marketed that are dangerous and have harmed people." There's concern, however, that the administration is underresourced and will only be able to adopt a staged approach to enforcement.[26] Certainly, if ex-FDA Commissioner Scott Gottlieb is to be taken at his word, things are about to change. His recent statement was unequivocal:

> There are a small number of unscrupulous actors who have seized on the clinical promise of regenerative medicine, while exploiting the uncertainty, in order to make deceptive, and sometimes corrupt, assurances to patients based on unproven and, in some cases, dangerously dubious products. These dishonest actors exploit the sincere reports of the significant clinical potential of properly developed products as a way of deceiving patients and preying on the optimism of patients facing bad illnesses.[27]

Second, by defining the broad categories of standards required, regulators can steer practitioners toward the areas of concern, and where they hope to see improvement. Researchers may not yet be able to devise the most appropriate potency assay, but they do know that potency is critical—as are purity, identity, and nontoxicity. By defining the broad categories of standards required, regulators can (and are) steering practitioners towards the areas of concern, and where they hope to see improvement. Progress will come in many small steps, not in one giant leap. We need to make sure we focus on the areas of concern and incorporate best practice, as it emerges.

Third, we need more research. To state the obvious: we will only fill the gaps in our knowledge by imaginative, well-designed research projects.

Finally, we need patience. Clinicians must not use vulnerable patients as experimental tools. Companies must not put immediate profit ahead of developing sound business models. And patients, hard though it might be, must do their best to ignore the Sirens luring them onto the rocks.

9   Pluripotency

It is not often you see science transformed before your eyes. In 2006, the Annual Meeting of the International Society for Stem Cell Research was held in Toronto. I hadn't attended this meeting in previous years. The science had been impressive, and by day three I was saturated. That lunchtime, I was still deciding which of the tempting sessions to attend next, when a friend told me that a friend had told her that one particular talk was not to be missed. It was just luck therefore that I was in the audience for Shinya Yamanaka's lecture. When I left the lecture theater an hour later, like the rest of the audience, I was in no doubt that I'd just witnessed the transformation of stem cell science.

What Yamanaka had presented to the ISSCR was the biological basis for pluripotency. This ugly name disguises a beautifully simple concept: a pluripotential cell is one that can make all the cell types that comprise the body.[1] I used the term "multipotential" in chapter 2 to describe the capacity of neural stem cells to generate neurons and glia, a term we similarly applied to bone marrow stem cells that can make all blood cell types. By comparison, a pluripotent stem cell can generate brain cells, blood, muscle, skin—all the cell types that make up the adult organism. A pluripotent cell is the ultimate stem cell.

This property may be all encompassing, but it is also ephemeral. As far as we know for certain, only a tiny cluster of cells during the whole of mammalian development are pluripotent, and they remain so for only a few hours or days. As the fertilized egg starts its developmental journey down the oviduct toward the uterus, it divides to form a ball of cells. Although, initially, all the cells in this cluster appear the same, by the time there are a few hundred cells, two populations have emerged—an outer ring of cells and an inner cluster. The outer ring will contribute to the placenta and the extraembryonic tissue. The inner cluster—called the "inner cell mass"—will form the embryo. These are the pluripotent cells, from which the entire body will be generated.

It is hard to exaggerate the importance of the concept of pluripotency for embryologists. In chapter 3, we marveled at the capacity of the multipotential stem cells of the fetal nervous system to generate an entire adult brain structure, apparently unassisted. But pluripotent stem cells are more powerful still. The obvious question is: what is it about them that gives them this enormous capacity? Naturally, this is a question biochemists and cell biologists might hope to address, but two problems arise. The inner cell mass is tiny, perhaps a tenth the size of the period at the end of this sentence. It is also ephemeral. Within a few days, it has compartmentalized itself further to generate the cell layers from which the fetus will emerge—the ectoderm, mesoderm, and endoderm—and pluripotency has disappeared. This doesn't give a scientist much to work with.

What researchers required was a stable population of cells that would permanently retain their pluripotency while they were grown in the laboratory. The first significant breakthrough came from embryonal carcinoma (EC) cells.[2] Teratomas are a group of particularly nasty cancers. They form tumors composed of multiple different tissues all confused in a malignant mass. A typical teratoma might include muscle, brain, gut, even teeth, but in an unstructured morass. It makes sense that pluripotent cells might underlie such growths, and sure enough, if teratoma cells are cultured, lines of pluripotent cells emerge. Being pluripotent, these EC lines can generate many different cell types. Indeed, if they are injected into a mouse embryo, they join the inner cell mass and contribute to different tissues as the embryo develops.

This discovery of EC cells was exciting news for stem cell biologists trying to understand pluripotency, but it also attracted the attention of those interested in regenerative medicine. The argument is by now a familiar one: if these EC cells could generate all the cell types in the body, couldn't they regenerate lost tissue? As would be expected, EC cells can make neural cells, from which lines of neurons can be generated. One neuronal line, "NTN2," showed considerable promise for use in a cell replacement strategy in preclinical studies,[3] and even reached the clinic. Unsurprisingly, however, being derived from teratomas, these cells had accumulated chromosomal abnormalities and mutations. Consequently, even though preclinical evidence suggested the cells were safe, the disquiet that arose at the prospect of transplanting tumor-derived cells into the human brain killed off the project. It was hard to be confident that they wouldn't carry their tumor-forming potential with them. So although EC cells kept the biochemists busy for a few years, they were not the therapeutic breakthrough we were seeking.

Somehow, the inner cell mass cells themselves needed to be cultured so that they expanded in number without losing their pluripotency. In chapter 6, we noted that there are broadly two ways expansion might be achieved. Either genes could be expressed in cells that would drive them to divide, or the cells could be treated with factors that would bring about the same outcome. We saw examples of the former approach with the conditional immortalization of neural stem cells, but the key to expanding inner cell mass cells turned out to be the latter approach.

If the inner cell mass cells are dissociated and plated onto plastic tissue culture dishes, they simply die. If, however, the plastic dishes are first covered with fibroblasts—a cell type easy to grow and expand—and the inner cell mass cells are plated on top of this feeder cell layer, then they stick to the fibroblasts and start to grow. They'll expand to form a colony, which can then be divided and expanded further on fibroblasts, ultimately giving rise to a cell line of billions of cells—all derived from a single inner cell mass cell.

The pivotal question of course is, has pluripotentency been retained? And it transpires that it has. Again, this can be demonstrated by injecting the expanded cells into a mouse embryo—adding them to a developing inner cell mass. As this injected embryo develops, most of its cells come from its own inner cell mass, but the injected cultured cells also make a contribution. The mouse is, in other words, a chimera: an animal composed of two sources of cells. Thus the ephemeral pluripotency of the inner cell mass has been captured and transferred in perpetuity to lines of cells, called "embryonic stem cells" (ES cells), a term coined by Gail Martin.[4]

This technology has considerable significance for regenerative medicine. More immediately, it has enormously stimulated the study of genetics and embryology. The ES cells injected into the chimeric mouse contribute to all tissues of the mouse fetus, and this includes the germ cells that generate gametes. So some of the eggs or sperm that the chimeric mouse will go on to produce will be derived from the ES cells. Subsequently, by selective mating, a second generation can be bred entirely from the original ES cells. So, mice can be generated entirely from the ES cells.

This technology has considerably advanced the development of transgenic animals. Scientists who wish to study the function of a gene typically want to undertake two experiments. They want either to remove the gene—"knock it out" to use the scientific vernacular—or to turn up its expression. By studying the consequences of these "loss of function" or "gain of function" studies, they hope to work out what the gene normally does. Knocking out or overexpressing a gene in cells in culture is not too difficult, but the real challenge is to do this in a living animal. Various techniques had been developed to introduce DNA constructs directly into the embryo, but these were clumsy and inaccurate. Typically, researchers might end up with either a few copies of the introduced DNA in the injected embryo or thousands, hardly conducive to controlled experiments. With the advent of ES cell technology, however, the genome of the cells could be manipulated in culture with relative ease, and animals could be generated from these manipulated ES cells. It became possible to generate permanent strains of laboratory animals (typically mice) with knocked-out or mis-expressed genes.

More subtle experiments could also be devised. It became possible, for example, to put a human gene into a mouse, and ask whether it would work the same as the mouse gene, or to put a mutated gene into a mouse associated with a specific disease risk in humans, and ask what the consequences were for the health of the mouse. These transgenic strains have become immensely valuable in biomedical research, as increasingly nuanced genetic manipulations have become possible.

Human Pluripotent Cells

None of this mouse biology necessarily had to be applicable to humans. In fact, stem cell biologists at the time were repairing the reputational damage emanating from an earlier failure to transfer findings from experimental animals to humans.

"Animal cloning," as it came to be known in the popular media, had given us "Dolly the Sheep" and "Snuppy the Dog."[5] To clone an animal, stem cell biologists would start with a fertilized egg. They would remove and discard the nucleus of the egg, putting in its place a nucleus taken from a donor animal—a skin cell, for example. The egg would now carry the genome of the donor and thus would be a clone—that is, a genetically identical copy—of the donor animal. If the egg was implanted into a surrogate mother, a new baby animal would be born, an exact genetic copy of the donor. In this way, in principle, the biologists could generate a whole tribe of animals genetically identical to the one founder animal.

This cloning technology worked well in sheep, dogs, and many other species, but for reasons that still aren't entirely clear was not transferable to humans. Unfortunately for the reputation of human embryology, in 2009 the South Korean scientist Woo-Suk Hwang claimed to have cloned human embryos in exactly this fashion. The subsequent disgrace of the fraudulent Hwang was accompanied by some soul-searching among stem cell biologists, and did nothing for the reputation of stem cell science.[6]

In truth, cloning works very inefficiently even in species where it does work, which means researchers need to start with a good source of eggs. Quite apart from the ethical alert over cloning human babies, a concern with the Hwang procedure was where he was getting his human eggs. It tranpired that many were donated by his junior female colleagues. Since the procedure required the women to be injected with hormones to make them superovulate then subject themselves to surgical removal of the eggs, this looked suspiciously like coerced donation. We were invited by Hwang to consider that Korean women were simply more "public spirited" than their Western counterparts.[7]

The real leap forward in making human pluripotent cells came in disguise. As you'll recall from chapter 1, the biomedical breakthrough of 1978 was in vitro fertilization (IVF). Infertile couples were able to donate eggs and sperm, have them combined in a laboratory, then have a newly formed embryo introduced back into the uterus of the prospective mother, who (with luck) would be able to bring the embryo to term and give birth to a child. While most observers rejoiced at the birth of Louis Brown, the world's first "test tube baby," some bioethicists spotted a problem. To be sure there would be enough eggs to work with, the female donors had to be induced to superovulate. This was good news for the clinics performing the IVF procedure because it meant that they could choose the best, and freeze the rest. But what to do with the fertilized eggs that were surplus to requirements? These were potentially new human beings. Indeed, in the view of some people, they were already new human beings. If like the Catholic Church, you considered life to commence at conception, then these artificially created embryos were ethically equivalent to the rest of humanity.

Initially, the frozen embryos could be kept for the donating couples to have multiple attempts at pregnancy, if the first try failed. But what happened to the remaining eggs once success was achieved? Was it permissible to simply destroy these potential human beings? If not, should they be stored in perpetuity, and, if so, who should pay for their storage? Decades later, there is still no consensus on this issue. There are currently an estimated 600,000 frozen embryos in the United States that are never likely to be used but can't be legally destroyed, at least not using federal funds.[8] In the United Kingdom, they are destroyed, and all that is is required for is for the "parents" to fail to respond to a letter from the fertility clinic asking what they would like to happen next to the embryos.[9] Since there is a cost to keeping them, quite probably the embryos are destroyed when the prospective parents decide they no longer think they are worth the expense.

Bioethicists have squirmed with these problems ever since, but stem cell biologists had a ready answer: give the embryos to us! In so far as this has occurred, so has commenced the attempts to create human ES cells. This was precisely the material required to investigate whether the mouse biology was indeed applicable to human embryology. The answer, it transpired, was that compared to the mouse, generating human ES cells was superficially similar, yet fundamentally different, and remarkably slow to resolve.

The two species were superficially similar in that the key to growing human ES cells was co-culture with fibroblasts, exactly as had been the case with mouse cells. But the two were also fundamentally different because the role played by the fibroblasts in support of the ES cells was quite distinct in the two species.[10]

The key component produced by the feeder cells for the mouse cultures was "leukemia inhibitory factor" (LIF), a factor had already been identified by hematologists through its action in regulating the differentiation of blood cells. The LIF produced by fibroblasts in the feeder layer acts to hold the mouse stem cells in a pluripotent state. However, LIF is much less important to human ES cells, for which "fibroblast growth factor" (FGF) and "transforming growth factor-beta" (TGFβ), also secreted by fibroblasts, do the same job. Slightly surprisingly, the two sets of factors even work by different mechanisms. So, although mice and humans are both placental mammals, different pathways seem to have evolved to control pluripotency. This is not an enormous surprise: we've already noted that 75 million years of evolution separates humans from rodents. Unremarkably, therefore, rodent embryology differs from human in many regards. For example, rodents have a feature called "diapause"—the temporary developmental arrest that occurs in one litter of pups if the previous litter is still suckling. The different role of LIF in mice might be associated with this feature.[11] But note also that what we're considering here is not the maintenance of normal pluripotency in vivo, but the artificial creation of cell lines in culture, something possibly quite different.

Despite these biological differences, you might imagine that the superficial similarity—culture on fibroblasts—would have meant that human ES culture would follow quickly on the heels of the mouse work. In fact, seventeen years elapsed between the discovery of mouse ES cells in 1981 and James Thomson's description of the first successful culturing of human ES cells.[12] Indeed, the culturing of ES cells from nonhuman primates had to be mastered by Thomson's lab before the human method finally emerged.

Cracking the biology, however, still hasn't made the ethical problems go away: there is still no international consensus on the use of human ES cells. There isn't even consensus across the European Union. A 2013 review by the European Science Foundation found that EU countries varied between "very permissive" in relation to embryo research (Belgium, Sweden, and the United Kingdom) to "very restrictive" (Croatia, Germany, Italy, Lithuania, and Slovakia). In four EU countries (Austria, Ireland, Luxembourg, and Poland), there was effectively no legislation on human ES cell research at all.[13]

The UK authorities don't like to be considered permissive, but they pride themselves on having evolved a pragmatic approach to human stem cell biology. IVF began in Britain, and this provoked consideration of the ethics of embryo manipulation in the United Kingdom in advance of other countries. Pivotal was the Warnock Report to Parliament in 1982, which gave rise to the Human Fertilization and Embryology Act in 1990, updated in 2001 and again in 2008. The provisions of this act allowed for the creation and use of human ES cells subject to license and supervision. They also provided for a UK Stem Cell Bank to act as a repository of UK human ES cell lines and for lines from other countries, should they wish to deposit them there.[14]

UK law recognized a conundrum: that the human embryo has (in the words of the UK Medical Research Council's *Code of Practice for the Use of Human Stem Cell Lines*) "a special moral status," but nonetheless that "the embryo does not have the full rights of a person." Therefore, "while its creation nor its destruction are to be treated casually," that destruction is not forbidden under the law.[15] While this perspective has been vigorously challenged over the years, notably by organizations such as the Catholic Church and "right to life" groups, its adoption has been quietly accepted in Britain.

Other European countries, particularly those with a strong Catholic tradition, have reached a different conclusion. For example, German law—citing Article 1 of the Basic Law of the Federal Republic of Germany (the German constitution), which states explicitly: "Human dignity is inviolable"—unequivocally forbids the destruction of human embryos, while allowing the research use of lines generated outside Germany.[16]

Outside Europe, a similar degree of variability has emerged. Famously, under George W. Bush's presidency, the United States forbade the use of federal funds for the creation of human embryonic stem cells, though the use of certain existing cell lines was permitted, and the destruction of embryos using State or Private funds was not outlawed. "My position on these issues is shaped by deeply held beliefs. I also believe human life is a sacred gift from our creator." declared Bush.[17] Naturally, this compromise position was criticized by stem cell scientists eager to advance this exciting new technology, but it was also seen by many to be inconsistent, even hypocritical. How could it be unethical to generate human ES cells using federal funds, but ethical using state or private funds? Graeme Laurie, professor of medical jurisprudence in Edinburgh was just one of a number of commentators for whom this position made no sense: "It is a strange morality indeed that pins the moral status and life of the embryo on the question of who is paying for the research."[18]

In Asian jurisdictions, regulations are different again. China, for example, permits ES cell research under a code established in 2003. Yet individual proposals are overseen by local institutional review committees, giving rise to considerable variation in the authorization and surveillance that individual research projects receive. It has been suggested that ES cell research is consequently more widespread there than in the West.

How these different perspectives will ultimately impact the acceptance of regenerative medicines derived from ES cells remains undetermined. For example, the European Medicines Agency will eventually pronounce on novel ES cell–derived therapies for the European Union as a whole, but the extent to which these are adopted by member countries will hang on individual national legislation. It is not clear how this circle will be squared. I suspect the FDA will face similar problems in confronting a multifaceted constituency.

But to return to the biology, since 1998 human ES cells have been available for researchers to transform into cell therapies. Several have proven effective in preclinical studies of brain disorders, and a number are now very close to clinical trials. Before we consider that work, however, we need to understand why I was so excited by Shinya Yamanaka's lecture.

The Biological Basis of Pluripotency

Nuclear transplantation didn't start with Dolly the Sheep; it was an established protocol in a number of zoology labs by the 1960s. Through that decade and into the 1970s, John Gurdon and Ronald Laskey did a series of experiments culminating in a 1975 publication.[19] Strangely, though this paper has become pivotal in developing the concepts of both stem cells and pluripotency, neither term appears in the paper. In fact, the study primarily addressed an issue that to modern ears sounds slightly quaint.

DNA had been discovered and known to be the material of heredity long before the 1975 Gurdon, Laskey, and Reeves study, but one of the important unresolved issues was whether all cells retained a complete copy of the genome. We are so used now to accepting that every cell in the body contains two complete copies of all 20,000 human genes (one copy from each parent), that we forget that this had to be demonstrated. While it was clear by the 1970s that DNA held the genome, how genes were controlled was far from clear. We know now that a substantial proportion of the DNA encodes the regulatory sequences that control when and where each gene gets activated, so that brain cells switch on brain genes, liver cells switch on liver genes, and so forth. The first regulatory mechanisms began to be uncovered in the 1960s with the discovery by François Jacob and Jacques Monod of the "lac operon" in prokaryotes, but through into the 1970s, the idea remained that perhaps cells threw out the genes they didn't need. So, brain cells avoided turning on liver genes by permanently disabling them, or by discarding them entirely. Such an inelegant mechanism sounds implausible now, although something similar does in fact happen in lymphocytes as they determine which antibody or T cell receptor they are going to produce.

The Gurdon study addressed this issue using fertilized frog eggs, which just like fertilized mammal eggs, generate entire embryos, in this case to form tadpoles. The experiment was to replace the nucleus of the fertilized egg with a nucleus taken from a differentiated skin cell, then determine whether this engineered egg could still produce an embryo. If it could, then it demonstrated unequivocally that the skin cell had retained all the genes required to build an embryo. Had it failed, then the possibility would have remained that the skin cell had discarded or permanently inactivated some of the key genes. Needless to say, the experiment succeeded, and just as with Dolly, the egg had retained its potency despite the replacement of its nucleus.

For our purposes, however, the frog egg and Dolly experiments reveal something truly fundamental about cell potency. You can take a pluripotent cell, remove its nucleus and replace it with one recovered from a differentiated tissue, and behold, the cell retains its pluripotency.[20] So where does pluripotency reside? It must be in the cytoplasm. We are so used to think of the nucleus as the control center of the cell—we talk all the time of the DNA in the nucleus carrying a "blueprint" for creating an embryo—that we miss the significance of the cytoplasm. A blueprint is fine, but you need an architect to read it. So where does pluripotency reside? In the cytoplasm.

To be more specific, what the Gurdon and Dolly experiments show is that there must be factors in the cytoplasm that instruct the nucleus in the role that they jointly will play. The nucleus from the skin cell prior to transplantation would have been engaged in skin cell functions, but when transplanted into the egg under the influence of the egg cytoplasm it switched to the production of an embryo. The question this raises then is: what might these cytoplasmic factors be? The excitement of the Yamanaka lecture was quite simply that it provided the first empirical answer to that question. And not just that: it taught us how to make pluripotent cells at will.

The Yamanaka Experiment

The 2006 Yamanaka experiment was such an ambitious and elegant piece of cell biology that it serves as an example of biological science at its best: a clear question, a sophisticated approach, and exacting experiments conducted with accuracy and thoroughness.[21] And the true test: its pivotal findings were replicated precisely in many laboratories worldwide (including my own) with speed and remarkably little difficulty.

The question was simple: what were the cytoplasmic factors that made a cell pluripotent? Yamanaka and his junior collaborator, Kazutoshi Takahashi, began with an assay and a hunch. They guessed that the factors must be proteins, and they compiled a list of all the candidate proteins they could think of. Note: the factors needn't have all been proteins: there are many nonprotein regulatory molecules in a cell. So, they could have fallen flat on their faces right from the off. From previous investigations into pluripotency, they complied a list of twenty-four proteins that could plausibly be involved. There was no reason a priori to imagine that this list was complete, but the hunch was that somewhere among these twenty-four were hidden the factors they were after.

Then came the assay: they sought the factors required to turn a mouse fibroblast into a pluripotent cell. The problem was, how would they know if they had succeeded? We've seen already that pluripotent cells have demonstrable properties: they can generate chimeric mice, for example. But an assay needs to be fast and robust, and preferably conducted in a tissue culture dish or a test tube. The trick was to employ a "reporter gene." *Fbx15* is a gene that is usually switched off in fibroblasts yet switched on in pluripotent cells. So if they could activate the *Fbx15* gene in a fibroblast, they might surmise that it had been turned into a pluripotent cell. Not for certain, of course, but this was at least a first step. So, their lab generated a line of mouse fibroblasts engineered so that if the *Fbx15* gene were activated, the cells would stain blue.

Next came the heroically arduous stage. They created a viral vector for each of the 24 factors. Viral vectors are a standard means of artificially expressing a gene in a cell. So fibroblasts carrying one of these vectors would express the corresponding factor. Introducing all twenty-four vectors would cause them to express all twenty-four factors, and this is precisely what Takahashi and Yamanaka did. Although viral vectors were a well-established technology, this experiment was not for the fainthearted. To express twenty-four factors stably and robustly in one population of cells required perseverance and dedication, and, I suspect, it involved a good number of disappointing evenings nursing yet another failed experiment.

But it worked. They got fibroblasts to express all twenty-four factors, and, sure enough, the cells stained blue. Just maybe—assuming their *Fbx15* assay wasn't lying to them—hidden somewhere among the twenty-four candidate factors were the pluripotency factors they were seeking.

But which ones? Almost certainly all twenty-four weren't required, but which were important and which not? To disentangle them, Takahashi and Yamanaka repeated the multivector experiment another twenty-four times, each time now with twenty-three viruses, sequentially omitting one factor at a time. The logic was clear: if they left out a factor that was necessary for the *Fbx15* activation, then the cells would no longer stain blue. If the activation still worked with twenty-three, then the omitted factor could not be important. After testing each of the twenty-four using this "24-minus-1" strategy, they concluded that fourteen played no significant role, whereas omitting any of the remaining ten reduced the effectiveness of the *Fbx15* activation.

Was that it? Were these ten the factors they were looking for? Well, possibly, but there could be redundancy within the group of ten. To test whether they really needed all ten, they repeated the omission experiment, this time using a "10-minus-1" strategy, again leaving out one factor at a time. What finally emerged were four factors that seemed to be absolutely required and that together by themselves were sufficient to activate *Fbx15*. If any one of them was omitted, activation failed or at least worked much less efficiently. So they had it: four proteins (Oct3/4, *Sox2*, Klf4, and *c-Myc*) now generally referred to as "OSKM" or the "four Yamanaka factors" were all that were required to turn fibroblasts into pluripotent cells.

Except, of course, that wasn't what they had shown. All they'd shown was that these four factors could activate the *Fbx15* gene. But were the fibroblasts now truly pluripotent cells?

It transpired that indeed they were. In a further series of experiments, they showed that these reprogramed fibroblasts could form "embryoid bodies." These are clumps of cells comprising multiple differentiated cell types. They are not well organized, and they could never grow into a functioning fetus, but they do comprise derivatives of each of the three embryonic germ layers—ectoderm, endoderm, and mesoderm. Thus they show that the cells had this one pivotal feature of pluripotency.

Finally and quite remarkably, Yamanaka and Takahashi showed that if these cells were injected into a mouse embryo, they could even achieve the ultimate: a chimeric mouse where the reprogrammed fibroblasts contributed to the germ line of the mouse, so that subsequent breeding experiments generated lines of mice derived entirely from the reprogrammed cells. Just imagine: a mouse generated entirely from skin fibroblasts, whose only manipulation had been the transient expression of just four factors. Arthur C. Clarke was surely correct: there comes a point where advanced technologies are truly indistinguishable from magic.

Pluripotency Unleashed

Yamanaka called his reprogrammed fibroblasts "induced pluripotent stem cells" or "iPS cells." Several things quickly became clear following the 2006 publication. First, several other labs were close on Yamanaka's heels, and several similar papers followed. One came from the lab of James Thomson. Ironic that the man who first derived human ES cells and came within months of being the first to publish an iPS cell protocol, was not also a recipient of the Nobel Prize in Medicine or Physiology when it was awarded to John Gurdon and Shinya Yamanaka in 2012. Interestingly, the Thomson cocktail of genes was somewhat different from Yamanaka's, showing that the OKSM formula was not unique, and that there was some redundancy among the reprogramming factors. Each of the OSKM factors is representative of a different family of transcription factors. It turns out that individual members of these families can substitute for one another. So *Sox2* is the member of the *Sox* transcription gene family that Yamanaka identified, but other members of the family can do the job equally well. Since 2006, several different reprogramming strategies have emerged using other gene combinations, or more artificial approaches. Thus Takahashi and Yamanaka were probably somewhat less constrained at the outset than they imagined.

Within a year, human cells had been reprogrammed into iPS cells. This piece of molecular biology really did cross readily between species, and the OKSM factors worked equally effectively for human as for mouse. For regenerative medicine, this was the true breakthrough: iPS cells instantly became a starting point for the development of future cellular therapies.

There also followed considerable investigation into what was going on in the fibroblasts as they were being reprogrammed, and into whether the resulting iPS cells were really equivalent to ES cells. Broadly speaking, it appears that iPS and ES cells are "transcriptionally, epigenetically, and functionally equivalent,"[22] meaning they are alike in terms of the genes they express and their potential to generate other cell types.

The epigenetic equivalence is significant because it relates to what happens during the process of reprogramming. How a cell behaves is less a consequence of its genetic constitution than its epigenetic configuration. All the cells in our body have roughly the same genome, but each does different things because of the epigenetic modifications that each cell's genome carries. DNA and the proteins that bind to DNA are modified by the addition of chemical groups—methylation, acetylation, phosphorylation—which change how the DNA and protein pack together, how accessible they are to the cell, and how readily the genes get turned on or off. Each tissue—and each cell type within each tissue—has its own characteristic pattern of modifications, and this epigenetic formatting determines how the cell orchestrates its various functions. Ultimately, "pluripotency" is just a label for a particular strategy that a cell deploys, a strategy that results in an explicit cascade of cellular behaviors culminating in a certain outcome. So the challenge of reprogramming is to remodel the cell's epigenetic signature from that of a fibroblast, which results in the specific set of fibroblast-like behaviors, to that of an pluripotent cell. Reassuringly, the epigenetic makeup of iPS cells—that particular pattern of DNA and protein methylation, acetylation, and other modifications—mirrors that observed in ES cells pretty closely. So iPS cells aren't just mimicking pluripotency in some strange, artefactual way; they really have adopted the mantle.

The further point to emerge is that researchers don't need to start with fibroblasts. Probably any cell type will do. Many laboratories have started with blood cells, among them the California Institute of Regenerative Medicine, which has commissioned the production of a large bank of iPS cells starting from blood cells.[23] My lab starts with "keratinocytes" grown from plucked scalp hairs.[24] Another ingenious method starts with the few sloughed-off bladder cells that can be spun out of a urine sample.[25] So long as it is a living cell, you really can start with almost anything.

And you can use them to create almost anything. In principle, pluripotent cells can make all the cell types in the body. In practice, you need to work out precisely how to treat the cells to produce exactly what you want. Sorting this out might have been onerous except for the success of mouse embryology over the past thirty years in working out the signals and mechanisms that drive development. As a result, we have clear road maps leading from the fertilized egg to many of the most interesting cell types. We know many of the significant lineage decisions that cells make, when they make them, and what mechanisms influence those decisions.

So for example, if you want to make to make, say, a motor neuron, then you first have to direct the pluripotent cells to make ectoderm, rather than mesoderm or endoderm. Specifically, you need to tell them to make neurectoderm, the layer of cells that generates the brain and spinal cord. They need to be directed to make spinal cord, not brain: to make ventral spinal cord, not dorsal; to make motor neuron progenitors, not the progenitors that will give other types of spinal cord neurons; and, finally, to make motor neurons, not the other cell types (such as oligodendrocytes) that those progenitors are also primed to make. Of course, if you want to make a very specific population of motor neurons—ones that innervate extensor muscles, for example, rather than flexor muscles—then you have some further refinements to make.

Remarkably, many of these mouse mechanistic pathways worked well when applied to human iPS cells. Conseuently, within a decade of Yamanaka's discovery, an enormous range of cell types derived from human iPS cells were available in laboratories across the world. Many labs, for example, now routinely make a few billion human cortical neurons each week, with considerable consistency between preparations. So for many studies, we neuroscientists no longer need to kill a batch of mice to get brain cells, or to concern ourselves with whether these mouse cells are really behaving like human cells. We simply make batches of human brain cells—a resource never before available to biomedical science.

Stem Cells and Disease Etiology

ES and iPS cells have opened significant opportunities for cell therapies in the brain, a topic to which we'll return in the next chapter. Their impact on the study of disease mechanisms is outside of remit of this book, but is actually the area of greatest current impact on brain science, and worth noting in passing.

Researchers can now take cells from any person and generate iPS cell lines, which in turn can be differentiated into essentially any cell type of interest. This becomes a powerful tool for the study of human diversity. I can take your iPS cells, for example, and examine how they compare to my iPS cells. When your iPS cells make cortical neurons, do they follow the same developmental trajectory as mine?

I noted above that mammalian embryology had been a real success story in recent decades, revealing the mechanisms of development in considerable detail. Yet despite this success, we have learned little about what generates diversity. The reason, of course, is that, for the most part we have been studying rodent development. and experimentalists go out of their way to make their rodents as alike as possible. Why?—because variation is noise. The more each mouse differs from another, the less reproducible the data. Fine: but the irony for those of us who undertook neuroscience precisely because we wanted to understand why we each behave the way we do, is that we study a system from which that variety has been deliberately excluded.

iPS cells have reintroduced diversity, but in a way we can control. We can compare your cells to mine, or cells from two siblings, or cells from a patient suffering from a particular condition to those from a nonsufferer.

Where the approach is particularly powerful is the study of genetic disorders. We can take cells from patients with a genetic variant associated with a particular disorder—be it autism, Parkinson's disease, or migraine—and compare them to cells from individuals without that genetic risk factor, confident that the precise genomes for both groups have been captured in the iPS cells. Then, if we see any consistent difference between the behavior of the two sets of iPS cells, we can tentatively ascribe that to the genetic variation and propose (somewhat more tentatively) that this difference might be linked to that disorder.[26]

This approach is beginning to reap benefits across a number of CNS disorders, a good example being ALS.[27] This progressive neurodegenerative disease is caused by a loss of the motor neurons in the brain and spinal cord, the cells that control muscular activity. Cell death is accompanied by muscle wastage, paralysis, and finally death, usually within three to five years. ALS is primarily a sporadic disease with no clear cause, but roughly 10 percent of patients have a genetic form arising from mutations in one of a number of genes, several of which have been identified.

In a pivotal 2014 study, researchers at Harvard generated iPS cells from several of groups of ALS patients, some carrying mutations in *SOD1*, one of the more common genetic risk factors, but other patients with *C9orf72* and *FUS* variants, two other genes associated with the disease.[28] In each case, they were able to differentiate the iPS cells into motor neurons and compare the physiology of each patient's cells not only with one another, but also with iPS cells derived from controls—individuals with the unmutated form of these genes that most of us carry.

The conventional wisdom is that cell death in ALS is the result of the excitotoxicity we discussed in chapter 2 in relation to stroke, but this study was able to show that motor neurons from each patient group shared a different feature that distinguished them all from the controls. Each patient line showed a hyperexcitability; that is, the motor neurons from patient iPS cells were more likely than those from controls to fire action potentials in the absence of stimulation. This seemed to be the result of an overactive sodium current in the cells, which in turn seemed in some unexplained fashion to be the result of the abnormal folding of the mutated protein.

At this point in the study, the researchers were able to combine the iPS methodology with another novel piece of biotechnology—gene editing. The problem with interpreting the hyperexcitability result is that it could be explained by some other difference between patient and control cells. What if, in addition to the genetic mutation, the patient cells carried some other deficit? Perhaps the ALS patients had all been treated with some drug that had damaged all their cells, including the fibroblasts from which the iPS cells originated. In this case, the hyperexcitability might be the result of that other factor, and have nothing directly to do with the disease.

The DNA sequence in cells can now be modified with considerable accuracy and efficiency. Techniques have been available to edit genes for many years, but new developments have made them considerably more tractable. These allowed the researchers to repair the mutated form of the *SOD* gene providing the patient lines with the normal, unmutated variant. When they then differentiated these gene-edited patient cells into motor neurons, they discovered that the hyperexcitability had disappeared. So repairing the mutation removes the hyperexcitability, showing unequivocally that this property was indeed caused by the mutation, and not by some other fault lying hidden within the patient cells.[29]

This use of patient-derived iPS cells to reveal fundamental biological mechanisms associated with specific disorders is spreading across biomedical science. My own lab, for example, has used the same approach to reveal features of autism that had not been previously observed.[30] What makes this ALS study so exciting, however, is its significant impact on drug discovery.

The prospects for individuals suffering from ALS are grim indeed. Death almost inevitably follows within a few years of diagnosis. The usual treatment is the drug "riluzol," and that typically extends life just a few months. Novel treatments are in clinical trials, including the cellular therapy we considered in chapter 6. But the iPS-based study is exciting not only because of the novel disease mechanism it revealed, but because the cells themselves provided a good assay to look for new therapies. Could drugs be added to the cell cultures that would reverse their hyperexcitability, and if so, might they reverse the ALS pathology in patients?

We now know the answer to the first part of that question is yes, and we will soon know the answer to the second part. The Harvard researchers were able to show that the drug "retigabine"—already approved for the treatment of epilepsy—normalized the hyperexcitability. Mutant cells treated with retigabine showed levels of excitability comparable to those of control cells. This promising outcome has led to a clinical trial currently ongoing to investigate the potential of retigabine for the treatment of ALS.[31] Thus we have here a new way to model human disease using real human cells, and a new assay system to develop new therapeutics.

10 Histogenesis

Many advocates of cell therapies are of the opinion that pluripotent cells—ES and iPS cells—represent a significant breakthrough in the progress of regenerative medicine. Alan Trounson and Natalie DeWitt, writing in the "Science and Society" section of the journal *Nature*, describe their potential as "unique and extraordinary."[1] Not that there isn't also some skepticism. The claims for pluripotent cells, according to Theodore Friedman, a former chairman of the National Institute of Health's Recombinant DNA Advisory Committee "have been characterized by the kinds of exaggerations and elevated expectations that were seen in the field of gene therapy just a few years earlier."[2] Claims, he might have added, that are now broadly seen as the epitome of hype in the history of biomedicine.[3]

There's no question that deciphering the biological basis of pluripotency is an exquisite piece of experimental science, but so was the discovery of Pluto. Why should we conclude that new stem cell therapies will result? Trounson and DeWitt cite two parameters to support their enthusiasm: the scalability of pluripotent cells—the ability to expand them almost indefinitely—plus their pluripotency. Do these stand up to scrutiny?

The first of these is significant, though certainly not unique. We have met several examples already of expansive cell populations: ReNeuron's "CTX0E03" cells and SanBio's "SB623" cells, to name just two. Conversely, we have encountered instances where the inability to scale the production of an appropriate stem cell population severely limited its applicability, the most prominent being the fetal cells used for the Parkinson's trials. A major problem in that instance was the variability between the cells derived from aborted fetuses. One scalable source that could be applied to a series of patients would certainly represent success. We have also identified variability and lack of scalability as a problem with MSC therapies. So perhaps the point about pluripotent cells is not that they are uniquely scalable, but that they might allow us to scale up the cells we actually want, rather than forcing us to work with the cells that happen to be intrinsically scalable.

The ability to generate any cell type we want, at least in principle, is clearly enticing. Many of the cells that practioners would like to get their hands on are not readily available. We hear all the time of clinical practice restricted by the lack of availability of human tissue: only 10 percent of the demand for organ transplants is currently being met.[4] There are, for example, far more candidates for liver transplants than there are cadaveric livers available. What if we could just make liver cells starting from pluripotent cells? And, beyond that, there are therapies we can't even conceive of because the cells are simply not available. No one has ever been able to realistically contemplate a cell replacement therapy for motor neuron disease that actually started with human motor neurons because where would you start to source such cells? Not that such a therapy is available now, but step one has been completed: these cells have now been generated from pluripotent cells, giving the adventurous among the regenerative medicine community a substrate to work on.

There is, however, a third feature of pluripotent cells that distinguishes them fundamentally from what has gone before: their capacity for histogenesis. When we discussed multipotential neural stem cells in chapter 2, we highlighted their capacity to generate all the major cell types of the central nervous system (neurons, oligodendrocytes, and astrocytes). If we are now saying that the potential of pluripotent cells includes the generation of neural stem cells, which in turn have the potential to generate all the major cell types of the nervous system, you might ask: what has actually been achieved? We seem to have taken a circuitous route and arrived at the same place. There are two differences. First, starting with pluripotent cells gives us more control over precisely the type of neurons and glia we end up with. We'll see exactly how important this is when we reconsider Parkinson's disease. But, second, the neural stem cells derived from the pluripotent cells have a property that earlier generations of cultured neural cells never achieved: the capacity for histogenesis. Pluripotent cells can actually build tissue from scratch.

If you cultured the ReNeuron CTX0E03 cells appropriately, predictably they would give rise to neurons and glia. But they would be jumbled neurons and glia, with none of the structure that you would encounter if you looked at proper brain tissue. So while these cells were derived from human cerebral cortex, they have no capacity to organize themselves into something that resembles cortex. Compare this with the fetal progenitor cells from which they are derived. The cells of the fetal ventricular zone (which we met in chapter 3) build the adult cerebral cortex essentially single-handedly. Certainly, they need a blood supply to provide nutrients and an endocrine system to provide the right hormonal milieu, and they will eventually need to talk to other brain regions to ensure the right networks are created. It is also true that this region of the fetal brain incorporates cells from neighboring regions as development proceeds. But essentially, this population of ventricular cells builds the cortical structure unaided. To go back to the epigenetic concept that emerged in chapter 9, these cells have the correct genes primed to direct a cascade of behaviors that will ultimately lead their progeny to create human cerebral cortex in all its complexity. This is the wonder of development.

Remarkably, human pluripotent cells have this same capacity. If human ES or iPS cells are cultured appropriately, they make cerebral cortical precursor cells, which will attempt to build a cerebral cortex in the tissue culture dish. How far they get will depend on how they are handled. If they are grown in the conventional fashion as a thin layer coating the surface of a plastic dish, then they won't get very far. They will try to form a tubular structure equivalent to the neural tube they would form early in brain development, but on the flat surface of the dish this turns into individual flower shaped rosettes, as if the tube had been salami-sliced into a series of thin sections and laid side by side. Modest though this is, this constitutes histogenesis far beyond anything conventionally sourced neural stem cells ever achieve.

But to see the pluripotent derivatives at their best, you need a more permissive culture. If you allow them to float free, then they form true three-dimensional structures, and start to organize themselves into a rudimentary brain, populated with neuroepithelial cells. Then these neuroepithelial cells begin to generate neurons. Just as in vivo, they make the deep cortical neurons before they make the superficial neurons, so they build the cortex starting from the inside and growing outward, just as happens in normal development. Eventually, they build something that looks remarkably like the brain of a six-week-old fetus.[5]

But then they hit a snag. Without a blood supply, they can only go so far. No cell in the mature brain is more than a few cell diameters away from a blood vessel, and without the nutrients the blood provides, the cells can't survive. The little cortical organoids have no blood supply and they soon outgrow the capacity of the culture environment to meet their needs, bringing progress to a halt.

Overcoming this limitation is a major area of research in the regenerative medicine community. How far researchers get will play a large role in setting ambitions in this area for the next decade. If it turns out that pieces of cortex (or any tissue) can be grown to a reasonable degree of size and maturity, then the possibility arises (at least in theory) that such tissue pieces could be used to replace the areas of brain lost to stroke, for example. This would, however, need to be accomplished with a degree of consistency and accuracy considerably beyond what has currently been achieved. Quite apart from the technical challenge, the ethical and logistical problems facing such an approach are enormous. In truth, were this strategy to prove feasible, it would surely impact other somatic tissues well in advance of brain, and in fact progress across a broad range of tissues is progressing apace.[6] One could conceive of the transplantation of liver or pancreas organoids, for example. An exciting prospect indeed.

Clinical Applications

So where will these advantages of pluripotent cells lead in the arena of brain repair? There are three indications involving neural tissue where pluripotent cell-derived products are threatening to make an early impact: spinal cord injury, Parkinson's disease, and disorders of the retina. This is not to say that other indications are not also in play. The original drive to replace cells following stroke has switched to pluripotent cells, and encouraging preclinical findings have emerged there.[7] Similarly, therapies for Huntington's disease have also moved on to employ pluripotent cell derivatives.[8] Other CNS disorders seem likely to follow a similar path.

The most advanced, somewhat surprisingly, are therapies for spinal cord injury; "surprisingly" because the challenge in spinal cord injury does not appear to align with the strengths of pluripotent cells. We saw in chapter 6 that the primary problem in spinal cord injury is the rupture of spinal axons, a problem of axonal regrowth rather than replacement. A transection (partial or complete) of the spinal column breaks the nerve pathways running between the brain and the spinal cord, meaning the brain can no longer control motor activity below the break. Voluntary movement is lost.

In 2009, the FDA approved a clinical trial sponsored by Geron Corporation for a differentiated cell product (GNROPC1) derived from human ES cells for the treatment of spinal cord injury. The uneven progress of this trial presents a cautionary tale. The first patient was treated in 2010, only to have Geron terminate the study a year later with just four of the eight planned patients having been treated. This withdrawal was not a consequence of an anticipated failure of the trial, the Company maintained. True, the clinical results showed little sign of efficacy (preliminary though they were), but GNROPC1 had appeared safe, which is as much as can ever be reliably concluded from a phase 1 clinical trial. Rather, Geron put its decision down to financial concerns. By killing the study, the company claimed, they would save

$25 million, enough to fund half a dozen phase 2 clinical trials of their two cancer products.

There was, however, a suspicion that this was not the whole story. The scientific rationale for the Geron approach always looked somewhat tenuous. The product they developed (GNROPC1) was a preparation of oligodendrocyte progenitor cells (OPCs), differentiated from a human ES cell line procured from James Thomson, the original developer of the human ES cell technology. It was never clear why this particular population of neural progenitor cells should have a therapeutic effect in spinal cord injury. The cells' primary function, as their name suggests, is to produce oligodendrocytes. It was never clear how these myelinating cells of the nervous system should have any positive impact on lesioned spinal neurons. This is particularly true since the mature oligodendrocytes are known to inhibit axon regrowth and are thought to be one of the negative factors preventing spinal cord repair. Animal studies, however, suggested that there was functional recovery following engraftment of these progenitor cells, and this had opened the possibility of a clinical trial.

It is also the case that Geron had had an earlier scare when some of their treated rats developed spinal cysts. These were neither malignant nor a cause for concern, argued Geron, but the FDA put a hold on the study until more data could be provided. When these new data suggested that the cysts were the result of an epithelial cell contaminant, which was excluded from subsequent cell preparations, the trial was restarted. Not without some raised eyebrows, however, particularly since this was the first-in-human trial of a decidedly novel therapy, where more caution might have been expected. Audrey Chapman and Courtney Scala have argued that Geron's original decision to go ahead was flawed. In their view the preclinical data were inadequate, the regulators' decision-making process opaque, and the design of the study prevented a meaningful consent by patients.[9] No doubt Geron would point to the 22,000-page dossier it had had to submit to the FDA in order to obtain clinical trial approval as evidence that they'd walked the extra mile, and then some.[10] "Don't be the first one out the door" was the conclusion of Michael West, Geron's chief scientific officer (CSO), "The first one out the door gets all the arrows in his back."[11]

Either way, the Geron withdrawal looked like a serious early setback for ES-based therapies and the end of the road for Geron's particular application, but neither prediction has proven correct. In 2014, GNROPC1 (now renamed "AST-OPC1") reemerged in the hands of the California-based Asterias Biotherapeutics, Inc., which, on the back of a $14.3-million grant from the California Institute of Regenerative Medicine (CIRM), had acquired the Geron technology.

The Asterias trial has increased considerably in complexity. There are now five cohorts of patients either in progress or planned. In each case, the patients have suffered cervical level lesions, a variation from the earlier thoracic lesion studies. These clinical studies are ongoing and clinical data are still sparse. The first cohort of three patients showed no clinical improvement, though the second cohort of five patients (who received a larger dose of cells) has been reported to show some clinical progress. These are small numbers, naturally, and we will have to wait until the end of the study to know the true outcome. In the meantime, there are strong testimonials from the small number of individual patients who have seen dramatic improvement.[12]

Asterias has also completed a number of animal studies, building on the earlier preclinical data,[13] reporting both behavioral and pathology assessments. They have adopted a novel way of integrating the behavioral data into one overall score of the animal's gait. This has the advantage of lessening the uncertainty about which might be the most proper measure of locomotor performance, and of removing the suspicion that the investigators make lots of readings and simply report the most favorable. Using that combined parameter, rats with cervical spinal cord lesions showed improved overall scores following engraftment of AST-OPC1 cells after four months, compared with ungrafted animals.

Of the pathological outcomes, they saw reduced cavitation and increased myelination following engraftment of the cells. Somewhat like the cyst formation following stroke, cavitation following spinal cord injury is a consequence of the loss of neural tissue. Again, as in stroke, scar tissue forms and a space remains where previously there was neural tissue. A reduction is cavitation therefore represents success in terms of reducing the advancing pathology. The increased myelination follows more directly from the cell therapeutic of choice: oligodendrocyte progenitor cells would be expected to generate myelinating cells. Consequently, this program certainly looks more robust than it once did, but as ever, we await the pivotal clinical trial results to see the real outcome.

What difference has it made taking pluripotent cells as the starting material, compared to the earlier studies we considered? Have the attributes of pluripotency, scalability, and capacity for histogenesis made a demonstrable difference? Histogenesis is not much in evidence here. The ES cells have been employed primarily for their pluripotency, in this case their capacity to generate OPCs. The scalability of the ES cells, and indeed of the OPCs, has allowed the expansion of specific cell types into clinical production and is very much in evidence. Human OPCs are not a trivial cell type to acquire from primary brain tissue, and consequently little direct research has been done on this cell population. But the extensive research into rodent OPCs that has been conducted over the years has proved applicable to their human counterparts, facilitating this approach enormously.

What's still missing here, however, is a satisfactory manufacturing process. A major shortcoming of the fetal cell approach for Parkinson's disease was the uncontrolled mix of cell types that emerged from the fetal brain: progenitor cells, young neurons of multiple types, and various contaminating cells—vascular endothelial cells, fibroblasts, and others. While the AST-OPC1 cells are nowhere near as disparate, neither are they a wholly characterized population.

In the original study describing the preparation of the OPC1 cells, more than 95 percent of the differentiated cells were OPCs.[14] In their more recent publications, Asterias says that the proportion of OPCs in the clinical preparations is between 30 and 70 percent, with the identity of the remaining cells uncertain.[15] They label cells with a marker called "Nestin," a protein expressed in a wide variety of neural progenitor cells, but also found in a variety of other cell types during development. The cells do not label with markers that would identify neurons or recognized glial cell types. So, we can conclude that these interloper cells are probably neural, but of unknown phenotype.

This relatively uncharacterized and variable mix of cells is not really satisfactory. The sponsors of this trial have done the best they can, and as far as possible, they have shown the cells to be safe, at least in the preclinical studies. Nonetheless, you would hardly be reassured if you bought a prescription drug at the pharmacist to be told that the active ingredient is somewhere between 30 and 70 percent of the total material in the tablet, and the makers were somewhat uncertain as to what else was in there.

Achieving more consistency and reproducibility remains a major goal for the field, particularly as this Asterias case is actually one of the more satisfactory examples of current practice: at least they know the identity of half the their cells. In many cases, it is simply impossible from the published literature to discern the true make up of the cell populations that are going into clinical trials.

Consider this intriguing comparison: if an academic research group wished to publish their data describing the biological activity of a population of cells, they would submit their manuscript to an academic journal, who would send it for assessment to a number of peer reviewers, experts in the field. The first question these reviewers would address would be: what are the cells in question, and how do the research team know that these cells are what they think they are? If the team had not demonstrated to the satisfaction of the reviewers that they had exhaustively characterized and identified their cells, it is highly unlike that the paper would be accepted for publication, certainly not in any reputable journal. Remarkable then, that regulators are relatively relaxed about exactly this question when the population of cells in question is about to be injected into someone's head. Surely this cannot be allowed to go unchallenged for too much longer. We need to know what's in the syringe.

There is a genuine conundrum here. A regenerative medicine project usually begins with an experimentally defined cell preparation, typically improvised by academic researchers, who will characterize the cells, at least in part, and show that they have some activity in an animal or cellular model. If successful, this project then gets adopted by a commercial entity, usually a small company. They optimize the protocols and scale-up, aiming for a clinical trial. Before the cells progress far into the clinical program, the protocols will often be handed over again, sometimes to a newly created development arm of the original company, and sometimes to an external contract research organization (CRO) that specializes in commercial production. Finally, a cell preparation emerges that finds its way into increasing numbers of patients. But the conundrum is: how can the anyone be certain that the cells emerging from this end process are equivalent to the cells the academic researchers started with. Unless they can, how do they know that the cells going into patients have equivalent biological activity to those tested in the preclinical models?

One answer of course is that the late development cells can be tested in the same animal models in which the earlier cells had been shown to work, and the best research groups do indeed do this. The problem is that such models are invariably slow and expensive. What the company really needs are fast and easy release assays, so that each batch of cells can be shown to be comparable. Partly because of cost, partly because of the difficulty of determining the mode of action of cell therapies, these release assays are slow to emerge.

Parkinson's Disease

The second area where pluripotent cells spell progress is Parkinson's disease. Here the approach seems much more methodologically secure. When we left this area in chapter 5, it was marooned without a suitable supply of cells. There was proof of concept that the replacement of dopaminergic neurons could bring about clinical improvement, but the source of the neurons—human fetal cells—was inconsistent and unreliable. Perfect, it would seem, for a defined cell product derived from pluripotent cells. At least three groups worldwide have declared their intention to begin trials with such cells in the near future. A consortium comprising Malin Parmar from Lund and Agnete

Kirkeby from Copenhagen recently announced support from Novo Nordisk in a project aimed at generating new dopaminergic cell therapies.[16] In the United States, a parallel program is under way sponsored by BlueRock Therapeutics and driven by stem cell biologists Gordon Keller and Lorenz Studer with the support of Bayer.[17] And in Japan a group led by Jun Takahashi in Kyoto are taking a similar approach.[18]

This progress rests firmly on the earlier pioneering work coupled with insights that have emerged with the advances in stem cell science. Significantly, it also hangs on some important findings from basic developmental biology. It turns out that, from the outset, biologists had misunderstood the origin of the midbrain dopaminergic neurons. Chapter 3 described how neurons are derived from the neuroepithelial cells of the neural tube. Broadly that is true, but it now transpires that the dopamine progenitors come from a distinct structure—the floor plate—adjacent to the region originally thought to produce these cells. This probably explains in part the variability that was seen with the early fetal grafts, and the insight has provided a more accurate process of dopamine cell production and concomitant improvement in the markers by which that process can be monitored. Coupled with the scalability inherent in the pluripotent cell starting material, these developments have all led to a much more robust and efficient manufacturing process being devised for the production of the dopaminergic neuron progenitor cells required for Parkinson's therapy.[19]

Cell therapy for Parkinson's disease has also taken another important step. Several times in this book we've had cause to bemoan scientists' poor understanding of the mode of action of cell therapeutics. As we've seen, we do not know how a number of cell therapies actually work. This has never really been the case for the dopaminergic therapy, where the evidence clearly supported replacement of neurons as the significant factor. Nonetheless, this hypothesis still needed proof.

Lorenz Studer and his colleagues provided this proof in mice using an elegant new technology, optogenetics. They took mice that had been lesioned using the same model we met in chapter 5: dopamine neurons on one side of the brain are destroyed, and the animal rotates, chasing its tail. This parkinsonism is then "cured" by the engraftment of dopaminergic cells, in this case human dopaminergic neurons derived from pluripotent cells, the sort of cells that are now about to enter clinical trials. This is all as we've come to expect: the lesion gives the mice a parkinsonian pattern of behavior, and the cell transplant restores the behavior to normal. The question is: can we be sure that the dopaminergic activity of the cells is responsible for bringing about the change in behavior?

This is the same predicament that Brian Cummings and his team faced in 2005 with the spinal cord grafts that we considered in chapter 6; they overcame the problem with the judicious use of tetanus toxin. But in the intervening years, technology had moved on, and Studer's team were able to employ a much more sophisticated approach.

The trick was that the cells they used were not just ordinary ES cells. They had been engineered to express a strange light-sensitive protein called halorhodopsin. As its name suggests, this protein is related to the rhodopsin, which in the eye normally mediates the capture of light. But this chimeric protein is part light receptor and part membrane chloride pump, and has a particular property. If the chloride pump is activated in dopaminergic neurons, it inhibits their firing, and this activation can be achieved (courtesy of the halorhodopsin attachment) by simply shining light on it. So, if you shine light on the dopaminergic cells, they cease activity.

This allowed the experimenters to perform an elegant experiment. First, they could show that hemiparkinsonian mice, engrafted with the cells, did indeed recover. The dopaminergic cells had worked, reducing the parkinsonian behavior. Then, using a fiber optic light source fed directly into the mouse brain, the chloride pump was activated in the transplanted cells, turning off their activity. If the behavioral recovery were truly dependent on the dopaminergic activity of the grafted cells, this should lead to a loss of the recovery and the reemergence of the parkinsonian behavior, and, sure enough, this is precisely what the experimenters observed. This experiment proved conclusively that it was truly the dopaminergic activity of the transplanted neurons that was responsible for the functional recovery.

Though an enormously influential experiment in the context of cell therapies, it is by no means the only application of optogenetics to therapies more generally. There are many scenarios where the ability to turn neurons on or off can deliver novel therapeutic approaches. In the treatment of pain, for example, optogenetics are being investigated as a means to lessen the ability of pain neurons to fire. The use of gene therapy to deliver this type of treatment is very much in its infancy but will surely gain prominence in coming years.[20]

The very strong impression now is that in the age of pluripotent human stem cells and given the substantial improvements in understanding and manufacturing, the huge investment in cell therapies for Parkinson's disease might be about to pay off.

Retina

The treatment of retinal disorders—particularly *age-related macular degeneration* (ARMD)—is probably the most dynamic area of neural transplantation in the whole regenerative medicine arena. Several reports have emerged recently of well-structured clinical trials with different pluripotent cell–derived preparations. To read these reports is to be struck not only by the advantages conferred by the pluripotent cells, but equally by the advancements and sophistication that have emerged since the earlier studies.

Recall that the core problem with ARMD is the degeneration of the pigment epithelium—the dark-colored layer of cells that underlies the retina (see figure 6.2). Since this layer is required to maintain the viability of the photoreceptors of the retina, loss of epithelium results in degeneration of photoreceptors with a corresponding loss of vision. We saw in chapter 5 that efforts to replace the pigment epithelium with cells taken either from the periphery of the patient's own retina or from cadaveric retina produced inconclusive or problematic outcomes. Both sources of material are clearly limited and are themselves "old" pigment epithelium.

So the scalability and pluripotency of pluripotent stem cells are an enormous advantage here. Several protocols have been devised that produce pigment epithelium from human ES or iPS cells. Indeed, a recognizable manufacturing process for the production of a true advanced therapeutic medicinal product is now emerging. The study led by Lyndon da Cruz and Peter Coffey at the London Project to Cure Blindness incorporates a clear linear process to generate a defined product, which (subject to release criteria) is provided to a surgeon, who then employs a specifically designed delivery tool to engraft the tissue into patients. No doubt the sponsorship of Pfizer, while signaling a growing interest of Big Pharma in regenerative medicine, has also helped develop manufacturing capability. Whatever other skills pharmaceutical companies might bring, they certainly understand the manufacture of medicines.

Histogenesis is also starting to emerge as a feature. In the da Cruz and Coffey study and also in the California study reported by Amir Kashani and colleagues,[21] the medicinal product is not merely cells dissociated from a culture dish—as has been the case in most studies reported in this book—but rather a sheet of pigment epithelial cells, which da Cruz and Coffey call a "patch," prepared on a specifically engineered membrane. The Kashani product is prepared on the synthetic polymer "parlene," which has the advantage of being already established for use in a medical device. In both cases, the cells have tissue integrity as a result of having been grown on a synthetic structure. They have already formed a polarized epithelium sitting on a membrane equivalent to Bruch's membrane, the structure on which they would sit in the undamaged eye. This is important because the breakdown of Bruch's membrane is an integral part of ARMD pathology in the first place. So by combining the capacity of the ES cells to generate pigment epithelium together with an artificial membrane, these researchers have built a structure capable of truly replacing lost retinal tissue.

Not all have embraced this approach, and it does have its own problems. Steven Schwartz and his colleagues at the Geffen School of Medicine in Los Angeles have preferred a dissociated cell approach, pointing out that the surgery required to deliver a cell suspension to the virtual space behind the retina is considerably simpler than that required to deliver the structured implant. Another problem with the retinal patch is that the size of the delivery tool and the patch itself means that they can't be effectively tested in rodents, simply because they're too big. This led the London group to run a preclinical trial in pigs, whose eyes are closer to humans' in size and structure than those of rodents. They were able to demonstrate successful delivery of the patch in twenty animals in this fashion. Which approach is the safer and more efficacious will presumably emerge from the clinical trials. Furthermore, clearly progress across multiple fronts will be required to make these advanced therapies a success: in stem cell science, certainly, but also in surgical devices, materials design, and manufacturing process.

All three of the initiatives mentioned above are now in clinical trials. Schwartz and colleagues have reported the most advanced data from the largest group of patients. They've treated eighteen patients across four centers, separated into two patient cohorts. One cohort comprises older patients with the dry form of age-related macular degeneration, and the other, younger patients with Stargardt disease, a genetic juvenile form of macular degeneration. The protocol these researchers have employed (and which has been broadly adopted) is to treat the worse-affected eye, and to use the better eye as a control. The emerging safety data are good, though some patients have suffered complications from the surgery and others from the immunosuppression (about which more later). But by and large, there were no adverse events associated with the cellular therapy, and while numbers are still small, the efficacy data are very encouraging. An increase in pigmentation around the damaged area was observed in thirteen of the eighteen patients, suggesting the grafts had taken. More significantly, visual acuity demonstrably improved in nine ARMD patients and stabilized in a further three. The visual acuity outcome in the Stargardt patients was more modest, improving in three patients and stabilizing in a further three. Visual acuity was measured using a letter chart similar to one many of us have encountered at our local opticians. Each patient was asked to name letters of decreasing size. The smaller the letters a patient could distinguish, the better the patient's visual acuity.

With their more technically ambitious approaches, da Cruz/Coffey and the Kashani studies have fewer treated patients to report currently. Both studies showed that the patch had engrafted successfully, with evidence for functional pigment cells now underlying the retina. The London study reported two patients, and while the numbers are again small, the improvement is marked, with both patients able to read an increased number of letters on the chart—from 10 to 39 in one case, and 8 to 29 in the other—compared to before treatment. Kashani reports five patients. The improvement here appears more modest. Four of five patients showed no significant improvement, while the fifth could read 17 letters more than before the treatment. In both studies again, the safety data were good.

In all the cases just cited, the pluripotent cells of choice are human ES cells. What about the newer iPS cells? In Japan, Michiko Mandai and colleagues commenced a study with two separate trials in 2017, both using a sheet of pigment epithelium similar to those above but derived from iPS cells.[22] One study was autologous: the retinal cells came from iPS cells generated from skin fibroblasts donated by the patient himself. The other was allogeneic: the iPS cells came from an unrelated donor and were thus more directly comparable to the ES cell studies. Both of these trials, however, are currently on hold. There were to have been two patients in the autologous study, and five in the allogeneic study, but in each case only a single patient has been treated. While there is no reason to believe that treatment won't recommence in the future, the reasons for the halt are informative, and address issues around safety we have ignored up to now: genetic mutation and tissue rejection.

11 Mutation and Rejection

A sad fact of life for cell biologists is that they can't expand cells in culture without inducing mutations. Nor, for that matter, can the cells in our bodies grow without doing the same. Under an assault of environmental chemicals and radiation, our cells accumulate mutations. These "somatic mutations" impair tissue integrity and, in the worst cases, give rise to cancers. At least in the body we have an immune system doing its best to track down and remove the deviants. Cells in a culture dish have no such assistance. Worse, the conditions in culture encourage mutation. Imagine that, by chance, a single cell among ten million in a culture dish acquires a genetic change—a mutation—that improves its growth characteristics. As the cell population is expanded, the rogue cells will quickly outgrow their benign neighbors and, all things being equal, will eventually take over the culture. Imagine that this was a preparation of iPS cells taken from a patient with the intention of generating pigment epithelial cells to transplant back into the patient's retina. Would those epithelial cells be safe? Since their generation involves first growing billions of iPS cells then billions of pigment epithelial cells, all from a single skin fibroblast, there are more than enough cellular generations for mutations to arise over and over again. Quite simply, by the end of this expansion process, the cells will have accumulated some number of mutations. Worse, the number and identity of these variations will themselves vary: each time cells go through this process, the outcome will be different. And though in some cases, the mutations will be recognizably threatening because they have been seen in cancers before, many will be hard to assess.

To add to the complexity, the mutations can be on a variety of scales. They can be tiny, involving literally a single base pair in the chromosomal DNA. Though small, these "point mutations" can have a large effect because they mess with the genetic code, generating proteins with altered structure and disrupted function. But mutations can be much larger than this, involving tens, hundreds, or even millions of base pairs. Sometimes whole chunks of chromosome are lost, or turned back to front so they don't read properly. Sometimes bits of chromosome get duplicated so that cells carry an extra copy, and often the genes on that duplicated piece are active, thereby increasing the dose of that gene in the cell. These bigger mutations are called "copy number variations" and are often difficult to assess because they frequently involve many genes either duplicated or deleted, and so have complex functional outcomes. Cells in culture are known to accumulate the smaller point mutations, but they are also susceptible to the bigger copy number variations.

Cell biologists take steps to mitigate this risk but are unable to remove it entirely. If the cultured cells are stressed by overcrowding or lack of nutrients, then the growth of variant cells is encouraged. So culture conditions are monitored carefully and kept within precise limits. Nonethelss, mutations will still emerge, and studies have confirmed that human pluripotent cells are not spared these pressures, and moreover that some pretty unsavory mutations can arise.

P53 in a case in point. It is an important cancer control gene. Molecular biologists term it a "tumor suppressor" because it acts in multiple ways to restrict the emergence of cancers. It can activate DNA repair mechanisms, mending the damage caused by chemicals or radiation. It can prevent tumor cells from accelerating aound the cell cycle, and can kill them if they persist. So, having an intact *P53* is an important property for a cell. Of concern then that P53 mutations can accumulate in human pluripotent cells in culture. In a recent report, Florian Merkle and colleagues looked at 140 different human hES cell lines, including twenty-six that had been intended for clinical use, and found that five of these lines carried *P53* mutations.[1] Moreover, the mutations were of a type called "dominant negative," a type particularly associated with human cancers.

In one sense, mutations in P53 and other cancer genes are relatively easy to deal with, if not prevent. It is not difficult to design assays that would screen out mutations in *P53* and other cancer genes. Such assays are now available, and surely all responsible cell therapists in the future will employ them. If a batch of cells has accumulated *P53* mutations, you can simply discard them and start again—costly and irritating, but probably relatively infrequent judging from Merkle's numbers. The trick for cell manufacturers will be to improve their culture technology to minimize the mutation rate, and to refine their in-process screening to detect variants quickly.

 More difficult is dealing with the preponderance of mutations accumulated in genes with no tumor association. Many researchers, driven by the understandable desire to know more about their cultures, are beginning to sequence the whole genome of cells. With "next-generation sequencing," this is relatively cheap and quick. Unsurprisingly, sequencing reveals that many cultures carry genetic variants.[2] Some will be somatic, originating in the donor, but many will have arisen in culture. Most will be in genes whose function is poorly understood, and few will have any known association with cancer.

What to do with these data? There is a strong argument to say that this information is of limited value. Since every cell in every human body carries a number of harmless mutations, there's no reason to expect that those in transplanted cells are any less likely to be benign. The usual way to deal with the cancer risk is to run "tumorigenicity assays." These come in two forms: animal studies and culture assays. Culture assays assess whether the cells demonstrate particular growth characteristics such as being able to grow in the absence of a supporting substrate. Such "anchorage-independent" growth is a feature of cancer cells and thereby a means of identifying cells with a tumorigenic potential. More reliable, though less palatable, are animal study assays, in which an experimenter asks whether the cells can form a tumor in an animal, usually some unfortunate mouse. Strains of mice are now available that have a constitutively compromised immune system. Since these animals are particularly inept at combating cancer, the failure of a population of cells to form a tumor in such mice is accepted as pretty good evidence that they pose a low risk in human patients. Most of the therapies we have considered so far would have been subjected to such assays.

So, tumorigenicity assays have face validity. They seem to show whether cells are or are not tumorigenic. The genetic tests, however, have no such validity. Discarding cells with *P53* mutations would seem to make sense from a precautionary perspective. You might also argue that if the plan is to use the cells to make, say, pigment epithelium, you might want to screen important retinal genes for mutations, again just from a precautionary perspective. But that specific application apart, the genetic information adds little to the risk analysis. If you don't know what a given gene actually does, how do you know whether a mutation in that gene is dangerous? A future complication may be that as we learn more about genes and gene function so the pressure to over-interpret this sort of data will grow.

Returning to the Japanese autologous iPS cell trial, the reason given for the hold on the treatment of the second patient was that his iPS cells had accumulated genetic mutations in culture, and it was not considered safe to proceed.[3] Of particular concern were three copy number variations—deletions—that had arisen in the culture. One of the genes involved was indeed a gene previously shown to be associated with cancer.[4] Consequently, even though tumorigenicity assays were negative—the cells did not form tumors in immune-compromised mice—the three big genetic deletions was considered to be too big a risk and the trial was put on hold.

We can sympathize with those tasked with reviewing this trial, who, reports suggest, came under considerable pressure from many quarters in what was an enormously high-profile case in Japan.[5] The precautionary principle is clearly paramount with such a crucial, novel therapy. Nonetheless, future decisions will need to be more evidence-based if innovative therapies are to proceed.

We can sympathize more with the patient himself. This blind, elderly man had a skin biopsy taken to grow an autologous iPS cell line, with the expectation that, after a long wait, he would receive the retinal graft he needed. Instead, after waiting even longer while experts pondered the genetic results, he was finally told the operation could not proceed. If, as suggested, he was offered enrollment on the study's other trial, the one employing allogenic grafts—someone else's cells rather than his own—he might well have asked why he'd had to wait all this time for his own cells to come through if these other cells could do the job. The answer is a complex mix of risk assessment, scientific uncertainty, and pharmaceutical economics, but it starts with the problem of immunogenicity.

Tissue Rejection

The observation is well-established that tissue cannot simply be transplanted from one individual to another without the recipient mounting a rejection response to the foreign material, hence the careful tissue typing that accompanies blood transfusions and organ donations. The biology underlying this response is multifaceted, but is underpinned by the capacity of cells of the immune system to recognize invading cells as foreign by the proteins they express, and display these foreign proteins in such a way that the intruding cells can be attacked and killed. This immune response evolved, of course, to protect us against invading organisms, such as bacteria or viruses, but is unfortunately incapable of distinguishing dangerous invaders from benign therapeutic transplants.

Apart from a brief discussion in chapter 5, I've managed to ignore tissue rejection in this narrative so far for a very simple reason. The brain (and the retina) is an immune-privileged site. Evading immune surveillance, it can tolerate foreign antigens in a way most of the body cannot. For scientists trying to develop replacement cells for pancreas to treat diabetes, for example, immune rejection is a serious problem. The allogeneic approach we considered above for the retina would fail disastrously in the pancreas because iPS cell derivatives from another person's cells would be very quickly attacked and rejected.

That said, the immune privilege is only relative. In some of the trials we've considered, the patients would still have received immuno-suppression, without which they might well have rejected the transplanted cells. This is problematic because immunosuppressant drugs tend to be toxic and not well tolerated by patients. Often patients simply stop taking them because they so hate the side effects, clearly not an optimal situation.

For this reason, the autologous approach is attractive. If you are receiving your own cells, then your immune system is unlikely to see them as foreign. Moreover, this is an avenue opened up by iPSC technology. Clearly, an ES-based therapy has to be allogeneic, but in principle a personalized iPS line could be generated for every patient.

Except it couldn't. The cost of autologous iPS production is entirely prohibitive. Each therapeutic line would take the best part of a year to produce: the reprogramming of patients cells; the differentiation of the desired derivative; the safety testing on multiple different lines (because you cannot be sure in advance, which individual iPS clone will be suitable); then the production run to give the final therapeutic product. The cost has been estimated to be roughly $1 million per patient. Further innovations will reduce this figure, but surely never to the level of affordability required for broad use.

The Japanese study wanted to start with an autologous therapy to give themselves the best chance of success with this fledgling iPS cell technology. So the single patient who received the autologous retinal graft is currently unique and may remain that way. In the future, probably only billionaires need apply.

Which brings us back to the allogeneic approach and the question of how immune rejection might be overcome. Several approaches are currently under investigation, and which will work and which not is not yet clear. Tissue compatibility hangs on a complex set of cellular mechanisms, but central are a set of proteins that comprise the major histocompatibility complex (MHC). Although these proteins are also complex, each having many variable forms, just two sets of genes need concern us here: MHC class I and MHC class II. Between them, they mediate three pivotal cell interactions.

First, essentially all cells express MHC class I proteins on their surface. There are many variants of class I proteins, and which is expressed is determined genetically. So while every cell in my body expresses the same class I proteins (since they are all genetically identical), my class I proteins wil be different from yours. There is a type of immune cell called a killer T cell, whose job it is to cruise around the body, like some gestapo agent, checking everyone's class I identity. If it spots a cell carrying the wrong credentials, it induces cell death. A useful cell to have on your side if your objective is genetic purity.

The second category of interaction involves another assassin— the Natural Killer Cell. This cell seeks a different type of deviant. It identifies cells that are expressing no class I protein at all. This happens in tumor cells, which sometimes lose their class I expression, and may therefore escape the attention of other immune mechanisms.

The third category is more complex and involves MHC class II proteins. Class II proteins are not expressed by all cells, but by cells such as macrophages, whose job is to track down foreign cells or cellular debris. Such material is engulfed by the macrophage and digested. It then performs a strange operation. It presents small pieces of the foreign proteins it has ingested on its surface in conjunction with MHC Class II proteins. This antigen presentation attracts naïve T cells—T cells not yet dedicated to a specific immune target. As a consequence of this interaction with the antigen presenting cells, the T cells are primed to seek out material carrying those same antigens—other similar invaders, in other words—and attack those foreign cells.

These multiple categories of seek-and-destroy weapons present a formidable arsenal for transplanted allogeneic cells to overcome. How might stem cell science protect transplants from this attack?

There are several strategies currently in play to overcome this problem, though none has yet definitively proven its worth. The most obvious tactic is to do what hematologists have always done: seek a match. Clearly, the best match is your own tissue, the autologous option, but in the absence of that, an iPS line can be employed that closely matches your own MHC profile. The problem is: the numbers. Imagine we had to have a line to suit everyone. There are estimated to be more than 16,000 MHC variants occurring in combination. There would have to be thousands of iPS lines in order to cover each combination. Then from each of those would have to be derived the particular cell type required, be it retinal epithelium, dopaminergic neurons, or oligodendrocyte precursor cells. Each one of those lines would potentially be considered by regulators as a different medicinal product, and need to be tested as such. Clearly, this could ultimately run to millions of cell lines and is clearly not a manageable approach.

But perhaps it could be approached stepwise. First, although there are many MHC class I variants, there are only three dominant genetic loci, called "HLA-A," "HLA-B," and "HLA-DRB1."[6] So long as these three loci are matched, then organ transplant studies suggest that rejection can at least be attenuated.[7] There is, however, still a problem: most of us are heterozygous at these loci, that is, we'll have inherited a different variant from each parent. So most carry two genetic variants at each locus. This can be overcome by only using iPS cell lines whose donors are homozygous—where both maternal and paternal genes are identical. So, instead of having to match two variants per genetic locus, only one match is needed. Suddenly the numbers start to look fractionally more manageable. If the most common homozygous combination were used to make an iPS cell line, that line would potentially fit with 14.5 percent of the Caucasian population. The second most common combination would add a further 6.5 percent. So, by generating just the right two homozygous iPS cell lines, 20 percent of the Caucasian population become potentially matched recipients. Moving forward from there, however, with real ethnically mixed populations and decreasingly common variants, we meet a serious problem of limiting returns. You need 17 iPS lines to cover 50 percent of the European population, and this set doesn't travel well. Only 13 of the 20 most common European combinations appear in the 50 most common among Hispanic populations. This number drops to 8 for African-Americans, and just 3 for Asians. So again, the numbers quickly become unmanageable.

The other logistical problem with this approach is finding donors who are homozygous at these genetic loci. Marc Peschanski and colleagues have calculated that to find just a single individual carrying the most common Caucasian variant (the one that would fit 14.5 percent of the population), they would need to screen roughly 180 individuals.[8] To get one of each of the ten most common, they would need to screen 11,000 people. And it gets progressively worse from there on.

Despite this difficulty, projects are under way to develop "haplobanks," repositories of iPS cell lines suitable for allogeneic transplantation. Most notably, the Center for iPS Cell Research

Mutation and Rejection 221

and Application (CiRA) in Kyoto, Japan, is pursuing this alternative,[9] and an international collaboration, the Global Alliance for induced Pluripotent Stem Cell Therapies (GaiT), is seeking to coordinate these activities.[10]

Cost and Comparability

Quite apart from logistical problems, the haplobank strategy has two serious issues: cost and comparability. The cost problem is obvious: who's going to make the substantial financial investment required to establish and maintain these banks? Haplobanks have been compared to public umbilical cord blood banks such as exist in many countries, including the United States and the United Kingdom. But the difference is that cord blood banks have demonstrable public health utility, whereas haplobanks will need to overcome some major problems before such utility could be demonstrated.

The conundrum is this. The iPS cells are the starting material for a cell therapy product, rather than the product itself. The actual products are the retinal pigment epithelium cells, the dopaminergic neurons, or the oligodendrocyte progenitor cells derived from the iPS cells. Like all medicines, allogeneic cell therapy products have to be tested for quality, safety, and efficacy. Each product must complete preclinical and clinical evaluation. Thus each therapeutic cell line made from an individual iPS cell line would be a new cell therapy product. So a hypothetical haplobank of, say, twenty iPS cell lines, used to generate twenty retinal pigment epithelium cell lines would be deemed by regulators to have generated twenty distinct medicinal products, each one of which would need to go through the entire preclinical and clinical testing process. And if those same twenty iPS cell lines were then used to make, say, twenty oligodendrocyte progenitor cell lines, that would be another twenty distinct medicinal products.

Were this really required, it would threaten the entire project. The estimated cost to develop a single medicinal product has been estimated to be $2 billion.[11] Is any company seriously going to run this process twenty times (or even just twice) for what is effectively just a single medicine? Certainly, there would be economies of scale in running the processes in parallel, and each success would reduce both the risk and the cost. Nonetheless, this is surely not a viable prospect.

This is where comparability might help. Comparability is a well-established regulatory solution to a medicine manufacturing problem. Pharmaceutical companies frequently have to make a changes in their manufacturing process. Sometimes, the manufacturer can no longer find a supplier for a particular reagent required for the production of the drug, or a superior process is discovered for making the drug. Sometimes, the company just wants to transfer production to a new facility. Consistency is clearly important: every batch of a drug must be identical. So how can the company be sure that the revised product is unchanged despite the change in the process?

The answer is that regulators will ask for comparability studies to be conducted. The regulators and the manufacturer will have agreed on the medicine's "critical quality attributes" (CQAs), that is, the key properties that assure the quality, safety, and efficacy of the drug. The regulators will ask the manufacturer to demonstrate that the new version of the drug has the same CQAs as the original version, assuring confidence that consistency has been maintained.

One proposition is that the same concept of comparability could be applied to the range of iPS cell–derived products. Thereby, all twenty retinal pigment epithelium cell products derived from the twenty iPS cell lines could be considered comparable if one line had been tested and approved as a novel cell therapy product, and the other nineteen could be shown to have the same CQAs as the approved line and were therefore comparable. Instead of running the entire process twenty times, you run it just once, and show all the others are essentially the same.

Unfortunately, this proposal involves some wishful thinking. The first problem would be simply that comparability is not normally applied to starting materials. If a process starts from different materials, it would usually be deemed to be an altogether different process. A compliant regulator might allow this one to slip by, but other issues might be more sticky. For example, regulators are unlikely to be happy with the current level of variance between iPS cell lines. Currently, even two lines from the same donor can vary considerably in their growth and differentiation properties. The first challenge to comparability, therefore, will be to develop reprogramming strategies that are sufficiently robust and reproducible to generate cell lines that could conceivably be considered comparable. Beyond that, how can there be certainty that two cell lines will behave equivalently once injected into a patient, especially since they have been purposely chosen to be distinct in precisely the genes that determine how the body will react to the cell transplant?

Beyond these technical questions, there are conceptual problems enough to keep bioethicists engaged for a long while. Fairly obviously, for two products to be considered comparable, they would need to target the same patients. If a pair of cell lines have been specifically designed to be different precisely so that they can target two distinct patient groups—in this case, patient groups defined by their genetics—then surely they can't be considered comparable.

Even if the comparability concept is accepted by regulators, it may not get the haplobank approach over the line. Demonstrating comparability itself is a challenge, as Christopher Bravery has pointed out.[12] If the mode of action of a therapy isn't known (as is the case with many cell therapies), then how can it be satisfactorily demonstrated that two lines are functionally equivalent. Moreover, comparability does not completely alleviate the cost problem. Three batches of a product usually need to be analyzed for comparability to be demonstrated. This is itself a substantial undertaking. Then, even following registration, problems remain. All drugs, particularly new drugs, are subject to intense regulatory scrutiny. Regulators monitor therapies carefully to see if any adverse outcomes emerge once the therapies start to be prescribed by doctors out in the real world. If just one "comparable" product runs into a problem, how should regulators view the whole combined set of therapies? This could be particularly complicated since each haplotype cell line will be taken up at different rates, some genetic variants being much more common than others. Thus batch size and penetration rate for each variant are likely to differ widely.

As things currently stand, most of these issues remain unaddressed. Nonetheless, considerable sums of money are being invested in strategies that require comparability to be ultimately workable. Regulators, who contrary to the opinion of some generally do like to say "yes," are currently keeping relatively quiet. They may not be able to maintain that silence for too much longer.

Genome Editing

Is there a cleverer way out of this predicament? Couldn't we design a pluripotent cell that could act as a universal donor, suitable for all patients, regardless of their MHC profile? Well, gene editing provides one way this might be done. What if we simply took a single iPS lines and knocked out its MHC class I genes? The immune system would no longer see the cells as foreign, and that one line could be used for everyone. Except, as we've already seen, the immune system has evolved a strategy to counter this tactic, precisely because this trick is adoted by some cancer cells to evade surveillance. This is where the natural killer cells come in. Not only must a cell avoid showing the wrong class I markers; it must show the right ones to elude these killer cells.

So we need to be clever: we need to create a marker to keep the natural killers at bay without alerting the killer T cells, which are seeking out the wrong credentials. This is an area of intense research, and although no one has the final answer yet, an exciting recent approach is that of Germán Gornalusse and colleagues.[13] They engineered a decoy MHC into their iPS cells. First, they used gene editing to knock out a gene called "beta-2-microgulin" (*B2M*). Normally, *B2M* is required to partner the class I proteins on the cell surface. So by removing this gene, they were able to ensure that no native class I genes were expressed by the iPS cells. This kept the cells out of the clutches of the killer T cells, but what about the natural killers?

To fool them, the Gornalusse team engineered an artificial gene into the cells, composed of two parts: *B2M* coupled to a minor class 1 gene (*HLA-E*) that doesn't vary very much between individuals. On its own, *HLA-E* isn't sufficient to elicit a killer T cell response, but its appearance on the cell surface in conjunction with the engineered *B2M* is enough to elude the natural killers. By coupling the two partners together—*B2M* and *HLA-E*—the Gornalusse team ensured that the *B2M* couldn't partner any of the more immunogenic class I proteins, so these engineered iPS cells had the best of both worlds: evading the killer T's, while placating the Natural Killers.

The preclinical data accompanying this study are very encouraging, and the approach will probably end up being tested in the clinic. Still, because genome manipulation has become so easy, further refinements will surely follow. One obvious candidate is the insertion of a "suicide gene" into the cells. This would be the ultimate safety switch, so that if the transplanted cells turned rogue and became tumorogenic, a drug could be administered to the patient that would simply kill all the engrafted cells, and any progeny they had produced.

Dare I end this chapter with the prediction that neural cell therapies are finally poised to deliver? We have been here before. The ultimate success with advanced therapies always seems to be just around the corner. Nonetheless, there are reasons to be optimistic.

First, pluripotent stem cells really have liberated the field from the cell availability and scalability constraints that plagued early studies. Whether the therapies emerging from pluripotent cells prove to be safe and efficacious remains to be seen, but at least now the right cells will be available—reproducible and at scale—so they can be tested conclusively. The therapies may or may not work, but at least we should learn the answer, one way or the other.

A second cause for optimism is the advance in other parallel biotechnologies that are impacting cell therapies, permitting ever cleverer experiments. Studer's optogenetic study is just one such development. Technologists can add value to their cell products with refinements such as the HLA-engineered cells we considered above. As gene editing becomes increasingly tractable, pluripotent cells and their derivatives can be manipulated in increasingly sophisticated ways. Again, this is an area that is only beginning to be explored, but surely represents the future.

Nonetheless, there remains much to do. Skills beyond those of cell and molecular biology are beginning to be applied to neural cell therapies, yet there is still a long way still to travel before truly regenerative therapies for the brain are finally to emerge. We've seen that earlier brain cell replacement approaches were adopted with considerable naïveté: slurried suspensions of stem cells squirted into areas of brain damage. This won't do any longer. The "retinal patch" has taught us (if we hadn't noticed already) that design criteria need to be specified in advance, and products manufactured to those precise specifications. Certainly, this will lead to more products, like the retinal patch, a combination of cells and a supportive matrix. The search for the appropriate substrates has been progressing in parallel with stem cell science for some years. Natural decellularized materials were an obvious starting point, but synthetic biodegradable polymers have been combined with stem cell implants for a number of decades. In 2002, for example, Evan Snyder's group implanted neural stem cells into lesioned cortex on nanoparticles made from polyglycolic acid, and observed extensive growth of neural fibers into the host brain.[14] Recently, more elaborate substrates have been devised, like manganese dioxide (MnO_2) nanoscaffolds combining biological substrates such as "laminin."[15] Advances in bioprinting hold considerable promise in generating purpose-built tissues and organs, while nothing has yet been printed comes close to the complexity of complexity of true biology. How this forced simplification will impact function is currently unclear.

Moreover, identifying appropriate substrates is just step one. None of the artificial structures have as yet addressed serious clinical limitations, such as those associated with scaling. Any tissue of any reasonable size will require a blood supply. We noted in chapter 10 that while three-dimensional organoids can be grown from pluripotent cells in culture, the absence of a blood supply ultimately leads to metabolic failure. So, in addition to the complex histogenesis required to build the primary brain tissue, there needs to be a concomitant construction of blood vessels, with all their biological and mechanical complexities. Then follows the challenge of linking any newly formed vessels with the host blood stream. These are no simple tasks, and no therapeutic has come close to accomplishing them on a clinical scale.[16]

Conversely, there are opportunities that have scarcely been pursued. Gene therapies can be used in conjunction with cell therapies. In some nonbrain applications—immunotherapies in particular—this is proving to be highly effective. "Car-T cells," for example, are cancer therapies in which a patient's own T cells are removed, engineered to carry a gene that boosts the immune system's ability to identify and kill cancer cells, then injected back into the patient. The "cells plus genes" approach could be used in different ways. The cells could be engineered to release therapeutic molecules or factors that would improve their own survival and efficacy in the damaged brain. Safe to say, an enormous range of possibilities is starting to emerge.

12 Prospects

Chapter 11 ended on an optimistic note regarding the future of neural cell therapies. In this final chapter, I want to briefly consider an alternative outcome: that cell therapies might be superseded.

A recurring theme in this book has been that predicting the future in biomedical science is a particularly pointless task given how often breakthroughs appear apparently from nowhere. How many of us predicted the appearance of iPS cells, yet see how dramatically that development has changed the landscape. There are, however, a number of endeavors on the horizon that are likely to come of age in the next few years, and they might have the effect of pushing to one side the transplantation therapies we have considered in this book. Two seem to me jointly to point in a genuinely fresh direction. The first of these is direct reprogramming.

Direct Reprogramming

The work of Gurdon, Thomson, and Yamanaka revealed something quite remarkable: if a cell can be induced to express the appropriate factors, then its fate can be fundamentally transformed. In the case of iPS cells, terminally differentiated cells—from blood, skin, or endothelium—were reprogrammed into pluripotent cells: that is, from cells with the most restricted of fates to cells with the most expansive. This was a shock to conventional embryologists, who had come to consider certain developmental steps irreversible. It was believed by many that once cells had been channeled during early development into one of the three primary germ layers (ectoderm, mesoderm, endoderm) then that step could not be reversed. Reprogramming destroyed that argument, but it raised an even more provocative question: if the correct genetic formula could be found was there any cell transplantation that could *not* be engineered?

The technique of iPS cell reprogramming takes a differentiated cell backward in development. From there, the cell can move forward again from the pluripotent state to become any of the various differentiated progeny to which such a cell would normally give rise (figure 12.1). The new question was: could reprogramming move a differentiated cell sideways; to another differentiated cell, for example, or a progenitor cell with a different fate? Could a fibroblast be turned directly into a neuron or a muscle cell? Or could it be turned into a neural progenitor cell or a bone marrow stem cell?

Remarkably, the answer to all of these questions turns out to be yes. As ever in science, there were straws in the wind long before biologists realized this was truly the case. Long before Yamanaka, a team in Seattle had shown that fibroblasts could be turned into muscle cells with a single gene.[1] The gene in question, *MyoD*, we now know to be a member of a group of transcription factors (*bHLH* genes) intimately involved in cell fate decisions in diverse tissues—heart, muscle, and brain. At the time, however, the molecular control of cell fate was largely unknown, and the existence of families of transcription factors was only starting to emerge as a

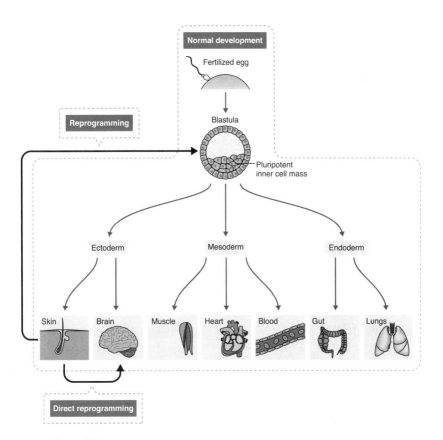

Figure 12.1

Pluripotency and reprogramming. During normal embryogenesis, the fertilized egg gives rise to a ball of cells called the "blastula," within which is a cluster of pluripotent stem cells called the "inner cell mass." These pluripotent cells generate all the cell types that make up the body. Reprogramming is the process whereby differentiated cells, such as skin fibroblasts, can be turned back into pluripotent cells. Direct reprogramming turns cells of one differentiated type (such as fibroblasts) directly into another (for example, neurons).

consequence of the early genome sequencing efforts. Colleagues, I recall, found this fate switch a troubling finding, but consoled themselves with the thought that these two cell types—fibroblasts and muscle cells—were actually pretty close embryologically, and anyway, strange things sometimes happened in tissue culture.

We have already met this phenomenon, "transdifferentiation"— the switching of cell fates—and noted that it has had a colorful history. While there were clear examples in vivo of cells apparently jumping from one fate to another, these were largely limited to "lower vertebrates" and involved closely related lineages. So, for example, if the limb of an amphibian is severed, cells within the stump dedifferentiate into progenitor cells (the "blastema"), which then regenerates multiple different cell types—muscle, dermis, bone—and thereby reconstitutes the lost tissue. In some species, heart cells (cardiomyocytes) can also dedifferentiate in response to damage, then redifferentiate following expansion to replace the heart tissue, and similar jumps have been observed in various tissues.[2] But these naturally occurring reprogramming episodes did not necessarily suggest that unrestricted reprogramming might be achievable experimentally.

Following Yamanaka, however, a simple formulation emerged. If the combination of factors that prescribed a particular fate could be identified, then quite plausibly, expressing those factors robustly might make a cell adopt that fate. While the extreme form of this theory probably doesn't hold up—that anything can be transformed into anything—nonetheless several quite remarkable steps have been demonstrated experimentally. Among them is the generation of neurons directly from fibroblasts.

The first demonstration of this came from Marius Wernig's laboratory at Stanford.[3] Their experiment reflected directly the approach that Yamanaka had pioneered. They sought the combination of transcription factors that would convert mouse skin fibroblasts directly into neurons, They found it required just three genes (*Ascl1*, *Brn2*, and *Myt1l*), and from this conversion emerged cells with all the significant properties of neurons: they grew a neuronal morphology, expressed the proteins that neurons express, formed synapses, and were electrically active. This was not, however, the first time that neurons had been directly reprogrammed from nonneuronal cells. Magdalena Götz and her collaborators had shown that transcription factors such as *Pax6* and *Olig2* modulated the capacity of glial cells to generate neurons.[4] But generating neurons directly from skin fibroblasts was an enormous leap in embryological terms: from a mesodermal end state (the fibroblast) directly into an ectodermal end state (the neuron), with no stem cell, or progenitor phase in between.

The neurons generated from this initial Wernig study, impressive though they were, were only characterized as generic neurons: no particular neuronal fate had been specified. The question therefore arose of whether specific populations of neurons could be generated. As we've seen, if the history of brain cell replacement has taught us anything, it is that we need the precisely correct neuron for each job. Several labs have now derived reprogramming formulas to generate specific neuronal populations, a number of which we've discussed in this book. For example, Ernest Arenas and colleagues at the Karolinska Institute in Stockholm have developed a protocol to generate dopaminergic neurons,[5] while Andrew Woo and colleagues at Washington University in St. Louis have made striatal neurons directly from fibroblasts.[6]

As well as indicating that clinically relevant neuronal populations are possible with this technology, these studies add a further wrinkle. It transpires that to achieve an optimal outcome, more than transcription factors need to go into the mix. At several points in this narrative, we've implied that cell fate can be determined by the correct combination of transcription factors. But as our understanding of cellular control mechanisms improves, we have discovered further cell components that participate in these processes. One such is noncoding RNAs.

For many years following the discovery of the genetic code in 1961, molecular biologists thought that the only essential role of DNA was to encode genes, which in turn encode proteins. Slightly alarming therefore was the discovery that only 1 percent or so of chromosomal DNA actually encoded conventional genes. The question became then: what is the other 99 percent doing? No less a person than Francis Crick is credited with concluding that it was probably "little more than junk."[7] So the term "junk DNA" entered the molecular biologists' vocabulary. But, of course, this had to be wrong. Were we seriously suggesting that a cell carried megabase upon megabase of DNA for which it had no use? Rather than deceiving ourselves by calling that 99 percent "junk," we needed to discover what it was actually doing.

We now know that much of the genome (though still not all of it) encodes RNAs that *do not* encode proteins. These RNAs have a direct function, rather than just being vehicles for the transport of protein-coding information from the nucleus to the cytoplasm. That function, in many cases, is to regulate the cell's translational machinery. They change the efficiency with which proteins are produced: proteins, which they themselves do not encode. Unsurprisingly therefore, they influence cell fate decisions, and can thereby influence reprogramming. In both of the direct reprogramming steps just cited, noncoding RNAs added to the mix improve the efficiency of the reprogramming steps.

This direct reprogramming has proven of interest to potential cell therapists for fairly obvious reasons. Instead of the laborious process of generating iPS cells, then taking them through a relatively long, complex process of differentiation, fibroblasts can be turned into the desired neuronal type in a single leap. There are, however, two issues with this approach, one practical and the other theoretical.

The practical problem is that, without the stem cell intermediate step, the possibility of expanding the cell population is lost. Neurons, as we know, are postmitotic: they don't divide. With the iPS cell approach, each reprogrammed fibroblast gives rise to a line of iPS cells that can be infinitely expanded, ultimately giving rise to billions of neurons. But with direct reprogramming, each reprogrammed fibroblast gives rise to a single postmitotic neuron. This does not amount to many cells. A halfway house might be to reprogram from fibroblasts to neural progenitor cells, bypassing the iPS cell, but still giving rise to a dividing cell, which can itself then be expanded to give rise to many neurons. Strategies are now in place to pursue this route.[8]

The theoretical issue relates to the mechanism underlying the direct reprogramming. Reprogramming iPS-style makes some sort of embryological sense. You make a pluripotent cell, then allow it to differentiate following the various embryological steps it would have taken in vivo. Direct reprogramming, however, makes no embryological sense. Nothing in nature, as far as we know, ever turns directly from a fibroblast into a medium spiny striatal projection neuron. This raises a number of questions regarding the veracity of directly reprogrammed change. Certainly, the reprogrammed cells have properties appropriate to the fate they've adopted, but have they abandoned all the indigenous programming that led them to their original fibroblast fate? This largely comes down to the epigenetic question we discussed earlier, and is the subject of current research.[9]

In Vivo Reprogramming

Direct reprogramming has the potential to enhance considerably the production of appropriate cells for stem cell transplantation. In combination with the final development I want to consider, there is the potential to make stem cell transplantation totally redundant. It would be ironic if reprogramming technology, which has done so much to liberate cell therapy from the constraints of cell availability and scalability, were to make the whole cell transplantation field obsolete, but this final development has the potential to achieve exactly that.

All the reprogramming we've considered so far takes place in a tissue culture system. What if cells could be reprogrammed in the patient? While reprogramming was limited to the production of iPS cells, this was not a plausible prospect. Turning a skin fibroblast into an iPS cell while it remained in a patient's skin would have done more harm than good. Since iPS cells have the potential to form teratomas, an iPS cell in a patient's skin (or anywhere else in the patient's body) would quickly give rise to a horrible tumor. But direct reprogramming avoids that risk, and performing the reprogramming directly in the patient also potentially overcomes the expansion problem.

How might this work? Take a population of nonneuronal cells in the brain that had the potential to be reprogrammed into neurons. Why couldn't the reprogramming vectors be injected directly into the brain, so they could so they could reprogram the nonneuronal cells near the damage site directly into the cell type that were lost? There are now a number of preclinical studies that have started to explore this scenario. For example, in 2014, Magdalena Götz and colleagues in Munich looked at direct reprogramming in the mouse brain following a stab wound.[10] They used a sharp blade to induce trauma in the mouse cerebral cortex, which previous work had indicated would activate a number of nonneuronal cells to proliferate in response to the injury. They followed the wound with the injection of a gene therapy vector encoding two transcription factors (*Sox2* and *Ascl1*). Their earlier work had suggested that these two factors would be sufficient to reprogram particular glial precursor cells, known to be activated by the injury, and turn them into neurons. Sure enough, the activated glial precursor cells in the wounded tissue incorporated the vectors, expressed the encoded genes, and new neurons began to appear in the damaged mouse brain as a consequence of this direct reprogramming.

Once again, these first experiments did not seek to generate any specific neuronal type, and that would certainly need to be achieved before a serious attempt could be made at therapy. But as we've seen, this specificity problem is being pursued in culture experiments, and is likely to translate in vivo. The big leap, of course, will be from mouse to human. We've already seen how easily that can come unstuck. Nonetheless, direct reprogramming of glial cells into neurons actually in the patient's brain is now a distinct possibility.

In this book, I've tried to relate the story of how neural stem cell therapies have grown from a flimsy, tentative idea into a robust and ambitious clinical program. While no cell therapy has yet to be licensed for any brain disorder, several therapies have entered proper clinical trials, which means that quite soon we will discover which work and which do not: whether the scientific and technical innovations we've encountered in this book can bring real benefit to patients suffering from intractable neurodegenerative disorders.

If we wanted to identify a single measure of how far we've come, it would probably be this. Attending conferences on advanced therapies in years past, one used to be overwhelmed by the technical hurdles yet to be overcome. In the last few years, that has subtly changed. Now the question most often asked as we huddle around the conference coffee outlet is: how are we going to make these therapies affordable? Our big worry is that, even if the therapies are proven to work, we won't be able to produce them cheaply enough to make them available to all who need them. As a scientist, one senses the manufacturers and regulators now looking over your shoulder, but one also has the gratifying sensation of having handed the problem on.

Acknowledgments

I have discussed the topics in this book with countless people, mostly without them realizing the extent to which I was picking their brains. These colleagues at KCL, NIBSC, the MHRA, and beyond, are too many to enumerate, but I'm grateful to them all for sharing thoughts and ideas so freely. Particular thanks, though, to Sandrine Thuret for the many times she has patiently explained to me the intricacies of the hippocampus, and to Ian Rees, who unselfishly shared his encyclopedic knowledge of medicines regulation. I'm enormously grateful to colleagues who have read and commented on sections of this book: Christian Schneider, Tytus Murphy, Pete Coffey, Roger Barker, John Sinden, and Elsa Abranches. Special thanks to Jenny Buckland, who read and corrected a huge chunk of the text. Needless to say, the remaining errors are my own.

I am indebted to Neil Smith, my illustrator. I'm not sure quite how Neil produced something so finished from the incoherent scribbles I sent him, but I'm terrifically impressed that he did. Thanks also to Matt Browne at the MIT Press for his patience, and to Peter Forbes at City University for his guidance.

During much of the writing of this book, I was employed by the Medicines and Healthcare Products Regulatory Agency. They accept no responsibility for any of the opinions expressed in this book.

Neither this book, nor much beyond, would ever have been achieved without the love and support of my wife, Helen New. To her, and to Florence and Matthew, I owe more than I can express.

Notes

Chapter 1

1. Stroke Association "State of the Nation," February 2018, 12. https://www.stroke.org.uk/sites/default/files/state_of_the_nation_2018.pdf.

2. Alzheimer's Disease International, "World Alzheimer Reports 2010," 2. https://www.alz.co.uk/research/files/WorldAlzheimerReport2010ExecutiveSummary.pdf.

3. Karran, E., and Hardy, J., "A Critique of the Drug Discovery and Phase 3 Clinical Programs Targeting the Amyloid Hypothesis for Alzheimer Disease," *Annals of Neurology* 76, no. 2 (2014): 185–205; quotation on 185.

4. Dirnagl, U., and Macleod, M. R., "Stroke Research at a Road Block: The Streets from Adversity Should Be Paved with Meta-Mnalysis and Good Laboratory Practice," *British Journal of Pharmacology* 157, no. 7 (2009): 1154–1156; quotation on 1154.

5. Miller, J. T., Rahimi, S. Y., and Lee, M., "History of Infection Control and Its Contributions to the Development and Success of Brain Tumor Operations," *Neurosurgical Focus* 18, no. 4 (2005): e4: 1–5.; quotation on 1. https://pdfs.semanticscholar.org/e72c/80c3f1f039166420bfd37c7b1c2385ec3d89.pdf?_ga=2.22448507.1722337118.1563849311-1234003552.1563849311.

6. Chamberlain, G., "British Maternal Mortality in the 19th and Early 20th Centuries," *Journal of the Royal Society of Medicine* 99, no. 11 (2006): 559–563. https://journals.sagepub.com/doi/pdf/10.1177/014107680609901113.

7. Macchiarini, P., Jungebluth, P., Go, T., Asnaghi, M. A., Rees, L. E., Cogan, T. A., et al., "Clinical Transplantation of a Tissue-Engineered Airway," *Lancet* 372, no. 9655 (2008): 2023–2030. http://doi.org/10.1016/S0140-6736(08)61598-6.

8. See "The Macchiarini Case: Timeline," *Karolinska Institutet News*, posted June 26, 2018. https://ki.se/en/news/the-macchiarini-case-timeline.

9. Thomas B. Okarma, as quoted in Boseley, S., "Medical Marvels: Drugs Treat Symptoms. Stem Cells Can Cure You. One Day Soon, They May Even Stop Us Ageing," *Guardian* (US edition), January 29, 2009. http://www.guardian.co.uk/science/2009/jan/30/stemcells-genetics/print.

10. King, N. M. P., and Perrin, J., "Ethical Issues in Stem Cell Research and Therapy," *Stem Cell Research & Therapy* 5 (2014): 5–85.

11. Fukuyama, F., *Our Posthuman Future: Consequences of the Biotechnology Revolution* (2002; repr. New York: Farrar, Straus and Giroux, 2006).

12. "Dr. William Haseltine on Regenerative Medicine, Aging and Human Immortality," *Life Extension*, July 2002. http://www.lifeextension.com/Magazine/2002/7/report_haseltine/Page-02.

13. More women than men suffer strokes.

14. See Takahashi, K., and Yamanaka, S., "Induction of Pluripotent Stem Cells from Mouse Embryonic and Adult Fibroblast Cultures by Defined Factors," *Cell* 126, no. 4 (2006): 663–676. http://doi.org/10.1016/j.cell.2006.07.024.

Chapter 2

1. The second "p" in "apoptosis" is not pronounced, at least not in the United Kingdom.

2. See Takahashi, K., and Yamanaka, S., "Induction of Pluripotent Stem Cells," *Cell* 126, no. 4 (2006): 663–676. http://doi.org/10.1016/j.cell.2006.07.024.

3. Leiden: Bone Marrow Donors Worldwide Eurodonor Foundation, 2013. https://www.cordbloodcenter.hu/wp-content/uploads/2016/12/BMDW 2013.pdf.

4. Burt, R. K., Balabanov, R., Han, X., Sharrack, B., Morgan, A., Quigley, K., et al., "Association of Nonmyeloablative Hematopoietic Stem Cell Transplantation with Neurological Disability in Patients with Relapsing-Remitting Multiple Sclerosis," *Journal of the New England Medical Association* 313, no. 3 (2015): 275–284.

5. In recent years, human embryologists concerned with in vitro fertilization have in recent years adopted the term "competence" to refer to the suitability of donated embryos for implantation. This is a quite different meaning of the term from the older usage I am employing here.

6. Chang, H. H., Hemberg, M., Barahona, M., Ingber, D. E., and Huang, S., "Transcriptome-Wide Noise Controls Lineage Choice in Mammalian Progenitor Cells," *Nature* 453, no. 7194 (2008): 544–547. http://doi.org/10.1038/nature06965.

Chapter 3

1. Richard Nowakowski, as quoted in Rubin, B. P., "Changing Brains: The Emergence of the Field of Adult Neurogenesis," *BioSocieties* 4, no. 4 (2009): 407–424, quotation on 418.

2. See, for example, Altman, J., and Das, G. D., "Autoradiographic and Histological Evidence of Postnatal Hippocampal Neurogenesis in Rats," *Journal of Comparative Neurology* 124, no. 3 (1965): 319–336.

3. See Altman and Das, "Autographic and Histological Evidence."

4. Classically, the dentate gyrus is not considered part of the hippocampal formation, but functionally it most certainly is, and that is the manner in which the association is meant here.

5. See Altman and Das, "Autoradiographic and Histological Evidence," 324.

6. Weiss, P., "Neuronal Dynamics and Neuroplasmic Flow," in Schmitt, F. O., ed., *The Neurosciences: Second Study Program* (New York: Rockefeller

University Press, 1970), 2:841. See also Gross, C. G., "Neurogenesis in the Adult Brain: Death of Dogma," *Nature Reviews Neuroscience* 1, no. 1 (2000): 67–73.

7. Kaplan, M. S., "Environment Complexity Stimulates Visual Cortex Neurogenesis: Death of a Dogma and a Research Career," *Trends in Neurosciences* 24, no. 10 (2001): 617–620.

8. Paton, J. A., and Nottebohm, F. N., "Neurons Generated in the Adult Brain Are Recruited into Functional Circuits," *Science* 225, no. 4666 (1984): 1046–1048.

9. For a full review of the literature on memory and the formation of new neurons, see Deng, W., Aimone, J. B., and Gage, F. H., "New Neurons and New Memories: How Does Adult Hippocampal Neurogenesis Affect Learning and Memory?," *Nature Reviews Neuroscience* 11, no. 5 (2010): 339–350.

10. Frankland, P. W., "Neurogenic Evangelism: Comment on Urbach et al. (2013)," *Behavioral Neuroscience* 127 (2013): 126–129.

11. For a review of the literature on looking for same kinds of stem cells and new neurons in human hippocampus that are observed in rats and mice, see Bergmann, O., Spalding, K .L., and Frisén, J., "Adult Neurogenesis in Humans," *Cold Spring Harbor Perspectives in Biology* 7 (2015): a018994.

12. Eriksson, P. S., Perflivieva, E., Bjork-Eriksson, T., Alborn, A.-M., Nordborg, C., Peterson, D. A., and Gage, F. H., "Neurogenesis in the Adult Human Hippocampus," *Nature Medicine* 4, no. 11 (2016): 1313–1317.

13. For a review of mathematical modeling of hippocampal neurogenesis, see Aimone, J. B., and Gage, F. H., "Modeling New Neuron Function: A History of Using Computational Neuroscience to Study Adult Neurogenesis," *European Journal of Neuroscience* 33, no. 6 (2011): 1160–1169.

14. http://www.seedmagazine.com/content/article/the_reinvention_of_the_self/.

15. Eisch, A. J., Cameron, H. A., Encinas, J. M., Meltzer, L. A., Ming, G.-L., and Overstreet-Wadiche, L. S., "Adult Neurogenesis, Mental Health, and Mental Illness: Hope or Hype?," *Journal of Neuroscience* 28, no. 46 (2008): 11785–11791.

16. For a review of comparative vertebrate neurogenesis, see Grandel, H., and Brand, M., "Comparative Aspects of Adult Neural Stem Cell Activity in Vertebrates," *Development and Genes Evolution* 223, nos. 1–2 (2012): 131–147.

17. Kempermann, G., "New Neurons for 'Survival of the Fittest,'" *Nature Reviews Neuroscience* 13, no. 10 (2012): 727–736.

18. Some authors use the alternative term "subventricular zone," though this is better applied to the equivalent fetal structure.

19. Niu, W., Zang, T., Smith, D. K., Vue, T. Y., Zou, Y., Bachoo, R., et al., "SOX2 Reprograms Resident Astrocytes into Neural Progenitors in the Adult Brain," *Stem Cell Reports* 4, no. 5 (2015): 780–794.

20. Quinones-Hinojosa, A., Sanai, N., Soriano-Navarro, M., González-Perez, O., Mirzadeh, Z., Gil-Pirotin, S., et al., "Cellular Composition and Cytoarchitecture of the Adult Human Subventricular Zone: A Niche of Neural Stem Cells," *Journal of Comparative Neurology* 494, no. 3 (2006): 415–434.

21. Inta, D., Cameron, H. A., and Gass, P., "New Neurons in the Adult Striatum: From Rodents to Humans," *Trends in Neurosciences* 38, no. 9 (2015): 517–523.

Chapter 4

1. Santiago Ramón y Cajal, as quoted in Rubin, "Changing Brains," 410. http://doi.org/10.1017/S1745855209990330.

2. Arvidsson, A., Collin, T., Kirik, D., Kokaia, Z., and Lindvall, O., "Neuronal Replacement from Endogenous Precursors in the Adult Brain after Stroke," *Nature Medicine* 8, no. 9 (2002): 963–970. http://doi.org/10.1038/nm747.

3. Arvidsson et al., "Neuronal Replacement from Endogenous Precursors," 968.

4. Magavi, S. S., Leavitt, B. R., and Macklis, J. D., "Induction of Neurogenesis in the Neocortex of Adult Mice," *Nature* 405, no. 6789 (2000): 951–955. http://doi.org/10.1038/35016083.

5. Diaz, F., McKeehan, N., Kang, W., and Hébert, J. M., "Apoptosis of Glutamatergic Neurons Fails to Trigger a Neurogenic Response in the Adult Neocortex," *Journal of Neuroscience* 33, no. 15 (2013): 6278–6284. http://doi.org/10.1523/JNEUROSCI.5885-12.2013.

6. Arvidsson et al., "Neuronal Replacement from Endogenous Precursors," 967.

7. Liu, F., You, Y., Li, X., Ma, T., Nie, Y., Wei, B., et al., "Brain Injury Does Not Alter the Intrinsic Differentiation Potential of Adult Neuroblasts," *Journal of Neuroscience* 29, no. 16 (2009): 5075–5087. http://doi.org/10.1523/JNEUROSCI.0201-09.2009.

8. Doetsch, F., and Scharff, C., "Challenges for Brain Repair: Insights from Adult Neurogenesis in Birds and Mammals," *Brain Behavior and Evolution* 58, no. 5 (2001): 306–322.

9. Kordower, J. H., Olanow, C. W., Dodiya, H. B., Chu, Y., Beach, T. G., Adler, C. H., et al., "Disease Duration and the Integrity of the Nigrostriatal System in Parkinson's Disease," *Brain* 136, no. 8 (2013): 2419–2431. http://doi.org/10.1093/brain/awt192.

10. Also valuable would be a diagnostic that could detect the neuron loss before it became so extensive. For a discussion of prospects in this regard, see Berg, D., Lang, A. E., Postuma, R. B., Maetzler, W., Deutsch, G., Gasser, T., et al., "Changing the Research Criteria for the Diagnosis of Parkinson's Disease: Obstacles and Opportunities," *Lancet Neurology* 12, no. 5 (2013): 514–524. http://doi.org/10.1016/S1474-4422(13)70047-4.

11. Although cell transplantation studies actually date back to the nineteenth century, they have had a significant therapeutic history only since the 1970s. See Barker, R. A., Parmar, M., Kirkeby, A., Björklund, A., Thompson, L., and Brundin, P., "Are Stem Cell-Based Therapies for Parkinson's Disease Ready for the Clinic in 2016?," *Journal of Parkinson's Disease* 6, no. 1 (2016): 57–63.

12. Bolam, J. P., and Pissadaki, E. K., "Living on the Edge with Too Many Mouths to Feed: Why Dopamine Neurons Die," *Movement Disorders* 27, no. 12 (2012): 1478–1483. http://doi.org/10.1002/mds.25135.

13. The initial report was the subject of an enthusiastic editorial. See Moore, R. Y., "Parkinson's Disease—A New Therapy?," *New England Journal of Medicine* 316, no.14(1987): 872–873.

14. See Dunnett, S. B., Björklund, A., and Lindvall, O., "Cell Therapy in Parkinson's Disease—Stop or Go?," *Nature Reviews Neuroscience* 2, no. 5 (2001): 365–369.

15. See US Food and Drug Administration, Code of Federal Regulations, Title 21, revised as of April 1, 2018. http://www.accessdata.fda.gov/scripts/cdrh/cfdocs/cfcfr/CFRSearch.cfm?fr=3.2.

Chapter 5

1. Trends in Pharmacological Sciences. http://www.cell.com/trends/pharmacological-sciences/pdf/S0165-6147(00)01725-9.pdf.

2. Drug Discovery Today. http://www.sciencedirect.com/science/article/pii/S1359644601019341.

3. Dunnett, Björklund, and Lindvall, "Cell Therapy in Parkinson's Disease—Stop or Go?" *Nature Reviews Neuroscience* 2, no. 5 (May 2001): 365–369.

4. Lindvall, O., Brundin, P., Widner, H., Rehncrona, S., Gustavi, B., Frackowiak, R., et al., "Grafts of Fetal Dopamine Neurons Survive and Improve Motor Function in Parkinson's Disease," *Science* 247 (1990): 574–577.

5. Olanow, W. C., Goetz, C. G., Kordower, J. H., Stoessl, A. J., Sossi, V., Brin, M. F., et al., "A Double-blind Controlled Trial of Bilateral Fetal Nigral Transplantation in Parkinson's Disease," *Annals of Neurology* 54 (2003): 403–414.

6. Isacson, O., Bjorklund, L., and Sanchez-Pernaute, R., "Parkinson's Disease: Interpretations of Transplantation Study Are Erroneous," *Nature Neuroscience* 4, no. 6 (2001): 553.

7. Ledford, H. "U.S. Scientists Fear New Restrictions on Fetal-Tissue Research," January 4, 2017, *Nature* 569, no. 7758 (2019): n.p. http://www.nature.com/news/us-scientists-fear-new-restrictions-on-fetal-tissue-re

search-1.21254?elqTrackId=89d72db5fb5a47ae8a8599f3afd63b77&elq=4a1658637a7248478da5136065b1b960&elqaid=17865&elqat=1&elqCampaignId=10597.

8. Carlsson, T., Carta, M., Muñoz, A., Mattsson, B., Winkler, C., Kirik, D., and Björklund, A., "Impact of Grafted Serotonin and Dopamine Neurons on Development of L-DOPA-Induced Dyskinesias in Parkinsonian Rats Is Determined by the Extent of Dopamine Neuron Degeneration," *Brain* 132, no. 2 (2008): 319–335. http://doi.org/10.1093/brain/awn305.

9. See Makin, S., "Pathology: The Prion Principle," *Nature* 538, no. 7626 (2016): 513–516. http://www.nature.com/nature/journal/v538/n7626_supp/full/538S13a.html.

10. US National Library of Medicine, ClinicalTrials.gov, "TRANSEURO Open Label Transplant Study in Parkinson's Disease (TRANSEURO)," "Detailed Description." https://clinicaltrials.gov/ct2/show/record/NCT01898390?term=TRANSEURO&rank=1.

11. Barker et al., "Are Stem Cell-Based Therapies for Parkinson's Disease Ready?" http://doi.org/10.3233/JPD-160798.

12. Zuccato, C., Valenza, M., and Cattaneo, E., "Molecular Mechanisms and Potential Therapeutical Targets in Huntington's Disease," *Physiological Reviews* 90, no. 3 (2010): 905–981. http://doi.org/10.1152/physrev.00041.2009.

13. For a review of studies on replacement of medium spiny neurons, see Golas, M. M., and Sander, B., "Use of Human Stem Cells in Huntington Disease Modeling and Translational Research," *Experimental Neurology* 278 (2016): 76–90. http://doi.org/10.1016/j.expneurol.2016.01.021.

Chapter 6

1. The exception to this generalization are the microglia, which are neither progenitor cells nor native brain cells, but modified blood-derived cells that colonize neural tissue.

2. La Manno, G., Gyllborg, D., Codeluppi, S., Nishimura, K., Salto, C., Zeisel, A., et al., "Molecular Diversity of Midbrain Development in

Mouse, Human, and Stem Cells," *Cell* 167, no. 2 (2016): 566–580; e19. http://doi.org/10.1016/j.cell.2016.09.027.

3. See, for example, Cattaneo, E., and McKay, R,. "Proliferation and Differentiation of Neuronal Stem Cells Regulated by Nerve Growth Factor," *Nature* 347, no. 6295 (1990): 762–765; Davis, A. A., and Temple, S., "A Self-Renewing Multipotential Stem Cell in Embryonic Rat Cerebral Cortex," *Nature* 372, no. 6503 (1994): 363–372. http://doi.org/10.1016/j.cell.2016.09.027; and Reynolds, B. A., and Weiss, S., "Generation of Neurons and Astrocytes from Isolated Cells of the Adult Mammalian Central Nervous System," *Science* 255, no. 5052 (1992): 1707–1710.

4. Reynolds and Weiss, "Generations of Neurons and Astrocytes."

5. See http://news.docckeck.com/en/366/stemcells-therapy-thanks-to-a-chopstick-accident/.

6. Zhu, J., Zhou, L., and XingWu, F., "Tracking Neural Stem Cells in Patients with Brain Trauma," *New England Journal of Medicine* 355, no. 22 (2006): 2376–2378. http://doi.org/10.1056/NEJMc055304.

7. Mossman, K., "The World's First Neural Stem Cell Transplant," *Scientific American*, December 18, 2006. https://www.scientificamerican.com/article/the-worlds-first-neural-s/.

8. Cummings, B. J., Uchida, N., Tanaki, S. J., Salazar, D. L., Hooshmand, M., Summers, R., et al., "Human Neural Stem Cells Differentiate and Promote Locomotor Recovery in Spinal Cord–Injured Mice," *Proceedings of the National Academy of Sciences* 102, no. 39 (2005): 14069–14074.

9. Kevin McCormack, as quoted in Keshavan, M., "With Its Stem Cell Therapies a Bust, Pioneering California Company to Shut Down," *STAT*, June 1, 2016. https://www.statnews.com/2016/06/01/stem-cell-company-shutters/.

10. Feldman, E. L., Boulis, N. M., Hur, J., Johe, K., Rutkove, S. B., Federici, T., et al., "Intraspinal Neural Stem Cell Transplantation in Amyotrophic Lateral Sclerosis: Phase 1 Trial Outcomes," *Annals of Neurology* 75, no. 3 (2014): 363–373. http://doi.org/10.1002/ana.24113.

11. See, for example, Treumer, F., Bunse, A., Klatt, C., and Roider, J., "Autologous Retinal Pigment Epithelium–Choroid Sheet Transplantation in Age Related Macular Degeneration: Morphological and Functional Results," *British Journal of Ophthalmology* 91, no. 3 (2007): 349–353. http://doi.org/10.1136/bjo.2006.102152.

12. Anderson, A. J., and Cummings, B. J., "Achieving Informed Consent for Cellular Therapies: A Preclinical Translational Research Perspective on Regulations versus a Dose of Reality," *Journal of Law, Medicine & Ethics* 44, no. 3 (2016): 394–401. http://doi.org/10.1177/1073110516667937.

13. See Azvolinsky, A., "Stem Cell Trials Data Mostly Go Unpublished," *Scientist*, May 5, 2017. http://www.the-scientist.com/?articles.view/articleNo/49350/title/Stem-Cell-Trial-Data-Mostly-Go-Unpublished/.

14. See Fung, M., Yuan, Y., Atkins, H., Shi, Q., and Bubela, T., "Responsible Translation of Stem Cell Research: An Assessment of Clinical Trial Registration and Publications," *Stem Cell Reports* 8, no. 5 (2017): 1190–1201.

15. See International Society for Stem Cell Research (ISSCR), *Guidelines for Stem Cell Research and Clinical Translation* (Skokie, Ill.: ISSCR, May 16, 2016). http://www.isscr.org/docs/default-source/guidelines/isscr-guidelines-for-stem-cell-research-and-clinical-translation.pdf?sfvrsn=2.

Chapter 7

1. Whitfield. J., Littlewood, T., Evan, G. I., and Soucek, L., "The Estrogen Receptor Fusion System in Mouse Models: A Reversible Switch," Cold Spring Harbor *Protocols* (2015). http://doi.org/10.1101/pdb.top069815.

2. Hicks, C., Stevanato, L., Stroemer, R. P., Tang, E., Richardson, S., and Sinden, J. D., "In Vivo and in Vitro Characterization of the Angiogenic Effect of CTX0E03 Human Neural Stem Cells," *Cell Transplantation* 22, no. 9 (2013): 1541–1552. http://doi.org/10.3727/096368912X657936.

3. See http://www.reneuron.com/clinical-trials/clinical-trials-overview/.

4. Steinbeck, J. A., and Studer, L., "Moving Stem Cells to the Clinic: Potential and Limitations for Brain Repair," *Neuron* 86, no. 1 (2015): 187–206; quotation on 192.. http://doi.org/10.1016/j.neuron.2015.03.002.

5. "Reports Positive Results in Phase II Stroke Trial," ReNeuron, December 5, 2016. http://4965zs3ha2l125fk78zkozo3.wpengine.netdna-cdn.com/wp-content/uploads/ReNeuron-PISCES-II-data.pdf.

6. See Smith, E. J., Stroemer, R. P., Gorenkova, N., Nakajima, M., Crum, W. R., Tang, E., et al., "Implantation Site and Lesion Topology Determine Efficacy of a Human Neural Stem Cell Line in a Rat Model of Chronic Stroke," *Stem Cells* 30, no. 4 (2012): 785–796. http://doi.org/10.1002/stem.1024.

7. Roberts, T. J., Price, J., Williams, S. C. R., and Modo, M., "Pharmacological MRI of Stem Cell Transplants in the 3-Nitroproprionic Acid–Damaged Striatum," *Neuroscience* 144, no. 1 (2007): 100–109.

8. Hicks, C., Stevanato, L., Stroemer, R. P., Tang, E., Richardson, S., and Sinden, J. D., "In Vivo and In Vitro Characterization of the Angiogenic Effect of CTX0E03 Human Neural Stem Cells," *Cell Transplantation*, 22, no. 9 (2013): 1541–1552. http://doi.org/10.3727/096368912X657936.

9. Hassani, Z., O'Reilly, J., Pearse, Y., Stroemer, P., Tang, E., Sinden, J., et al., "Human Neural Progenitor Cell Engraftment Increases Neurogenesis and Microglial Recruitment in the Brain of Rats with Stroke," *PLoS ONE* 7, no. 11 (2012): e50444. http://doi.org/10.1371/journal.pone.0050444.g006.

Chapter 8

1. Collins, F., "Regenerative Medicine: The Promise and Peril," *NIH Director's Blog*, March 28, 2017. https://directorsblog.nih.gov/2017/03/28/regenerative-medicine-the-promise-and-peril/.

2. Davey, M., "Stem Cell Therapies: Medical Experts Call for Strict International Rules," *Guardian* (US edition), July 6, 2017. https://www.theguardian.com/science/2017/jul/06/stem-cell-therapies-medical-experts-call-for-strict-international-rules.

3. Berkowitz, A. L., Miller, M. B., Mir, S. A., Cagney, D., Chavakula, V., Guleria, I., et al., "Glioproliferative Lesion of the Spinal Cord as a Complication of 'Stem-Cell Tourism,'" *New England Journal of Medicine* 375, no. 2 (2016): 195–196. http://doi.org/10.1056/NEJMc1605534.

4. Burke, S. P., Henderson, A. D., and Lam, B. L., "Central Retinal Artery Occlusion and Cerebral Infarction Following Stem Cell Injection for Baldness," *Journal of Neuro-Ophthalmology* 37, no. 2 (2017): 216–221.

5. See Wu Medical Center (WMC) website: http://www.wumedicalcenter.com.

6. Turner, L., and Knoepfler, P., "Selling Stem Cells in the U.S.A.: Assessing the Direct-to-Consumer Industry," *Stem Cell* 19, no. 2 (2016): 1–4. http://doi.org/10.1016/j.stem.2016.06.007.

7. Turner, L., "ClinicalTrials.gov, Stem Cells and 'Pay-to-Participate' Clinical Studies," *Regenerative Medicine* 12, no. 6 (2017): 705–719. http://doi.org/10.2217/rme-2017-0015.

8. Alliance for Regenerative Medicine (ARM), *2016 Annual Data Report* (Washington, D.C.: ARM, 2017). https://alliancerm.org/sites/default/files/ARM_2016_Annual_Data_Report_Web_FINAL.pdf.

9. Fung, M., Yuan, Y., Atkins, H., Shi, Q., and Bubela, T., "Responsible Translation of Stem Cell Research: An Assessment of Clinical Trial Registration and Publications," *Stem Cell Reports* 8, no. 5 (2017): 1190–1201. http://doi.org/10.1016/j.stemcr.2017.03.013.

10. Squillaro, T., Peluso, G., and Galderisi, U., "Clinical Trials with Mesenchymal Stem Cells: An Update," *Cell Transplantation* 25, no. 5 (2016): 829–848. http://doi.org/10.3727/096368915X689622.

11. Anderson, A. J., and Cummings, B. J., "Achieving Informed Consent for Cellular Therapies: A Preclinical Translational Research Perspective on Regulations versus a Dose of Reality," *Journal of Law, Medicine & Ethics* 44 (2016): 394–401.

12. See European Biopharmaceutical Enterprises (EBC), "EBC Position Paper on the Hospital Exemption" (2013). https://ec.europa.eu/health//sites/health/files/files/advtherapies/2013_05_pc_atmp/22_1_pc_atmp_2013.pdf.

13. See "U.S. v. Regenerative Sciences, LLC," 741 3d 1314 (2014), Leagle.com. https://www.leagle.com/decision/infco20140204141.

14. Epstein, R. A., "The FDA's Misguided Regulation of Stem-Cell Procedures: How Administrative Overreach Blocks Medical Innovation," Manhattan Institute (MI), September 24, 2013. https://www.manhattan-institute.org/html/fdas-misguided-regulation-stem-cell-procedures-how-administrative-overreach-blocks-medical-5897.

15. Gottlieb, S., and Klasmeier, C., "The FDA Wants to Regulate Your Cells," August 7, 2012, *Wall Street Journal*, Opinion, June 2019. https://www.wsj.com/articles/SB10000872396390444405804577558992030043820.

16. Chirba, M. A., and Garfield, S. M., "FDA Oversight of Autologous Stem Cell Therapies: Legitimate Regulation of Drugs and Devices or Groundless Interference with the Practice of Medicine?," *Journal of Health and Biomedical Law* 7 (2011): 233–272; quotation on 272.

17. Philippidis, A., "Stem Cell Tourism Hardly a Vacation," *Genetic Engineering & Biotechnology News* (GEN), August 16, 2012. https://www.genengnews.com/insights/stem-cell-tourism-hardly-a-vacation/.

18. Chirba and Garfield, "FDA Oversight of Autologous Stem Cell Therapies," 270–271.

19. Weijer, C., Shapiro, S. H., and Cranley, K., "Clinical Equipoise and Not the Uncertainty Principle Is the Moral Underpinning of the Randomised Controlled Trial," *British Medical Journal* 321, no. 7203 (2000): 756–757.

20. Swift, T. L., "Sham Surgery Trial Controls: Perspectives of Patients and Their Relatives," *Journal of Empirical Research on Human Research Ethics* 7, no. 3 (2012): 15–28. http://doi.org/10.1525/jer.2012.7.3.15.

21. See International Society for Stem Cell Research (ISSCR), A Closer Look at Stem Cells (website): www.closerlookatstemcells.org.

22. Dobkin, B. H., Curt, A., and Guest, J., "Cellular Transplants in China: Observational Study from the Largest Human Experiment in Chronic Spinal Cord Injury," *Neurorehabilitation and Neural Repair* 20, no. 1 (2006): 5–13; quotation on 5. http://doi.org/10.1177/1545968305284675.

23. "Nine Things to Know about Stem Cell Treatments," A Closer Look at Stem Cells. http://www.closerlookatstemcells.org/stem-cells-and-medicine/nine-things-to-know-about-stem-cell-treatment.

24. Davey, "Stem Cell Therapies," https://www.theguardian.com/science/2017/jul/06/stem-cell-therapies-medical-experts-call-for-strict-international-rules.

25. US Food and Drug Administration (FDA), "FDA Announces Comprehensive Regerative Medicine Framework" (news release), November 16, 2017. https://www.fda.gov/newsevents/newsroom/pressannouncements/ucm585345.htm.

26. Charo, R. A., and Sipp, D., "Rejuvenating Regenerative Medicine Regulation," *New England Journal of Medicine* 378, no. 6 (2018): 504–505. http://doi.org/10.1056/NEJMp1715736.

27. US Food and Drug Administrative (FDA), "Statement from FDA Commissioner Scott Gottlieb, M.D. on FDA's Comprehensive New Policy Approach to Facilitating the Development of Innovative Regenerative Medicine Products to Improve Human Health," November 16, 2017. https://www.fda.gov/NewsEvents/Newsroom/PressAnnouncements/ucm585342.htm.

Chapter 9

1. Takahashi and Yamanaka, "Induction of Pluripotent Stem Cells."

2. See Przyborski, S. A., Christie, V. B., Hayman, M. W., Steward, R., and Horrocks, G. M., "Human Embryonal Carcinoma Stem Cells: Models of Embryonic Development in Humans," *Stem Cells and Development* 13, no. 4 (2004): 1–9.

3. Watson, D. J., Longhi, L., Lee, E. B., T, F. C., Fujimoto, S., Royo, N. C., et al., "Genetically Modified NT2N Human Neuronal Cells Mediate Long-Term Gene Expression as CNS Grafts in Vivo and Improve Functional Cognitive Outcome Following Experimental Traumatic Brain Injury," *Journal of Neuropathology and Experimental Neurology* 62, no. 4 (2003): 368–380.

4. Martin, G. R., "Isolation of a Pluripotent Cell Line from Early Mouse Embryos Cultured in Medium Conditioned by Teratocarcinoma Stem Cells," *Proceedings of the National Academy of Sciences* 78, no. 12 (1981): 7634–7638.

5. Kim, M. J., Oh, H. J., Kim, G. A., Setyawan, E. M. N., Cho, B. C., Lee, S. H., et al., "Birth of Clones of the World's First Cloned Dog," *Scientific Reports*, November 10, 2017. https://www.nature.com/articles/s41598-017-15328-2.pdf.

6. Check, E., and Cyranoski, D., "Korean Scandal Will Have Global Fallout," *Nature* 438, no. 7071 (2005): 1056-1057. https://www.nature.com/articles/4381056a.

7. "Will the Regulator Please Stand Up? It's Time for the South Korean Government to Launch an Investigation into How Eggs Were Obtained for a Ground-Breaking Stem-Cell Experiment" (editorial), *Nature* 438, no. 7066 (2005): 257. https://www.nature.com/articles/438257a.

8. US Department of Health and Human Services, Office of Population Affairs (OPA), "Embryo Adoption," August 3, 2017. https://www.hhs.gov/opa/about-opa/embryo-adoption/index.html; Cauterucci, C., "What Should Be the Fate of a Spare Frozen Embryo?," *Slate*, January 20, 2016. http://www.slate.com/articles/double_x/doublex/2016/01/frozen_embryos_and_the_anti_abortion_activists_who_love_them.html.

9. Boseley, S., "Thousands of 'Neglected' Embryos Destroyed," *Guardian* (US edition), April 13, 2000. https://www.theguardian.com/uk/2000/apr/14/sarahboseley.

10. See Yu, J., and Thomson, J. A., "Pluripotent Stem Cell Lines," *Genes & Development* 22, no. 15 (2008): 1987–1997. http://doi.org/10.1101/gad.1689808.

11. Martello, G., and Smith, A., "The Nature of Embryonic Stem Cells," *Annual Review of Cell and Developmental Biology* 30, no. 1 (2014): 647–675. http://doi.org/10.1146/annurev-cellbio-100913-013116.

12. Thomson, J. A., Itskovitz-Eldor, J., Shapiro, S. S., Waknitz, M. A., Sweirgiel, J. J., Marshall, V. S., and Jones, J. M., "Embryonic Stem Cell

Lines Derived from Human Blastocysts," *Science* 282, no. 5391 (1998): 1145–1147.

13. Campos-Ruiz, V., *Human Stem Cell Research and Regenerative Medicine: Focus on European Policy and Scientific Contributions* (Strasbourg: European Science Foundation, 2013). http://archives.esf.org/fileadmin/Public_documents/Publications/HumanStemCellResearch.pdf.

14. Full disclosure: I am a former Director of the UK Stem Cell Bank.

15. UK Medical Research Council (MRC), *Code of Practice for the Use of Human Stem Cell Lines* (version 5) (London, MRC, April 2010), 10. https://www.mrc.ac.uk/documents/pdf/code-of-practice-for-the-use-of-human-stem-cell-lines/.

16. See Rolf, S., "Human Embryos and Human Dignity: Differing Presuppositions in Human Embryo Research in Germany and Great Britain," *Heythrop Journal* 53, no. 5 (2010): 742–754. http://doi.org/10.1111/j.1468-2265.2010.00601.x.

17. President Bush, as quoted in Park, A., "George W. Bush and the Stem Cell Research Funding Ban," *Time*, August 20, 2012. http://healthland.time.com/2012/08/21/legitimate-rape-todd-akin-and-other-politicians-who-confuse-science/slide/bush-bans-stem-cell-research/.

18. Graeme Laurie, as quoted in "Bush 'Out of Touch' on Stem Cells," *BBC News*, July 20, 2006. http://news.bbc.co.uk/1/hi/sci/tech/5197926.stm.

19. Gurdon, J. B., Laskey, R. A., and Reeves, O. R., "The Developmental Capacity of Nuclei Transplanted from Keratinized Skin Cells of Adult Frogs," *Journal of Embryology and Experimental Morphology* 34, no. 1 (1975): 93–112.

20. To be precise, the frog egg is "totipotent" rather than "pluripotent," but the distinction is not important in this discussion.

21. Takahashi, K., and Yamanaka, S., "Induction of Pluripotent Stem Cells from Mouse Embryonic and Adult Fibroblast Cultures by Defined Factors," *Cell*, 126, no. 4 (2006): 663–676. http://doi.org/10.1016/j.cell.2006.07.024

22. Martello and Smith, "Nature of Embryonic Stem Cells," 655. http://doi.org/10.1146/annurev-cellbio-100913-013116.

23. See https://www.cirm.ca.gov/our-progress/awards/generation-and-characterization-high-quality-footprint-free-human-induced.

24. Cocks, G., Curran, S., Gami, P., Uwanogho, D., Jeffries, A. R., Kathuria, A., et al., "The Utility of Patient Specific Induced Pluripotent Stem Cells for the Modelling of Autistic Spectrum Disorders," *Psychopharmacology* 231, no. 6 (2014): 1079–1088. http://doi.org/10.1007/s00213-013-3196-4.

25. Wang, L., Wang, L., Huang, W., Su, H., Xue, Y., Su, Z., et al., "Generation of Integration-Free Neural Progenitor Cells from Cells in Human Urine," *Nature Methods* 10, no. 1 (2012): 84–89. http://doi.org/10.1038/nmeth.2283.

26. See Dolmetsch, R., and Geschwind, D. H., "The Human Brain in a Dish: The Promise of iPSC-Derived Neurons," *Cell* 145, no. 6 (2011): 831–834. http://doi.org/10.1016/j.cell.2011.05.034.

27. Martinez, A., del Valle Palomo Ruiz, M., Perez, D. I., and Gil, C., "Drugs in Clinical Development for the Treatment of Amyotrophic Lateral Sclerosis," *Expert Opinion on Investigational Drugs* 26, no. 4 (2017): 403–414. http://doi.org/10.1080/13543784.2017.1302426.

28. Wainger, B. J., Kiskinis, E., Mellin, C., Wiskow, O., Han, S. S. W., Sandoe, J., et al., "Intrinsic Membrane Hyperexcitability of Amyotrophic Lateral Sclerosis Patient-Derived Motor Neurons," *Cell Reports* 7, no. 1 (2014): 1–11. http://doi.org/10.1016/j.celrep.2014.03.019.

29. Kiskinis, E., Sandoe, J., Williams, L. A., Boulting, G. L., Moccia, R., Wainger, B. J., et al., "Pathways Disrupted in Human ALS Motor Neurons Identified through Genetic Correction of Mutant SOD1," *Stem Cell* 14, no. 6 (2014): 781–795. http://doi.org/10.1016/j.stem.2014.03.004.

30. Kathuria, A., Nowosiad, P., Jagasia, R., Aigner, S., Taylor, R., Andreae, L. C., et al.,"Stem Cell–Derived Neurons from Autistic Individuals with SHANK3 Mutation Show Morphogenetic Abnormalities during Early

Development," *Molecular Psychiatry* 231, no. 6 (2017): 1079–1088. http://doi.org/10.1038/mp.2017.185.

31. US National Library of Medicine, ClinicalTrials.gov, "Clinical Trial of Ezogabine (Ratigabine) in ALS Subjects," February 2018. https://clinicaltrials.gov/ct2/show/NCT02450552?cond=ALS&intr=Retigabine&rank=1.

Chapter 10

1. Trounson, A., and DeWitt, N. D., "Pluripotent Stem Cells Progressing to the Clinic," *Nature Reviews Molecular Biology* 17, no. 3 (2016): 194–200; quotation on 194. http://doi.org/10.1038/nrm.2016.10.

2. Friedmann, T., "Lessons for the Stem Cell Discourse from the Gene Therapy Experience," *Perspectives in Biology and Medicine* 48, no. 4 (2005): 585–591; quotation on 585. http://doi.org/10.1353/pbm.2005.0089.

3. See, for example, Addison, C., "Spliced: Boundary-Work and the Establishment of Human Gene Therapy," *BioSocieties* 12, no. 2 (2016): 257–281. http://doi.org/10.1057/biosoc.2016.9.

4. Giwa, S., Lewis, J. K., Alvarez, L., Langer, R., Roth, A. E., Church, G. M., et al., "The Promise of Organ and Tissue Preservation to Transform Medicine," *Nature Biotechnology* 35, no. 6 (2017): 530–542. http://doi.org/10.1038/nbt.3889.

5. See Lancaster, M. A., Renner, M., Martin, C.-A., Wenzel, D., Bicknell, L. S., Hurles, M. E., et al., "Cerebral Organoids Model Human Brain Development and Microcephaly," *Nature* 501, no. 7467 (2013): 373–379. http://doi.org/10.1038/nature12517.

6. Clevers, H. "Modeling Development and Disease with Organoids," *Cell* 165, no. 7 (2016): 1586–1597. http://doi.org/10.1016/j.cell.2016.05.082.

7. Tornero, D., Wattananit, S., Gronning Madsen, M., Koch, P., Wood, J., Tatarishvili, J., et al., "Human Induced Pluripotent Stem Cell–Derived Cortical Neurons Integrate in Stroke-Injured Cortex and Improve Functional Recovery," *Brain* 136, no. 12 (2013): 3561–3577 http://doi.org/10.1093/brain/awt278.

8. Ma, L., Hu, B., Liu, Y., Vermilyea, S. C., Liu, H., Gao, L., et al., "Human Embryonic Stem Cell–Derived GABA Neurons Correct Locomotion Deficits in Quinolinic Acid–Lesioned Mice," *Stem Cell* 10, no. 4 (2012): 455–464. http://doi.org/10.1016/j.stem.2012.01.021.

9. Chapman, A. R., and Scala, C. C., "Evaluating the First-in-Human Clinical Trial of a Human Embryonic Stem Cell–Based Therapy," *Kennedy Institute of Ethics Journal* 22, no. 3 (2012): 243–261. http://doi.org/10.1353/ken.2012.0013.

10. Alper, J., "Geron Gets Green Light for Human Trial of ES Cell–Derived Product," *Nature Biotechnology* 27, no. 3 (2009): 213–214. https://www.nature.com/articles/nbt0309-213a.

11. Michael West, as quoted in Baker, M., "Stem-Cell Pioneer Bows Out: Geron Halts First-of-Its-Kind Clinical Trial for Spinal Therapy," *Nature* 479, no. 7374 (2011): 459. http://www.nature.com/news/stem-cell-pioneer-bows-out-1.9407.

12. See, for example, Aldrich, M., "Experimental Stem Cell Therapy Helps Paralyzed Man Regain Use of Arms and Hands," *USC News*, September 9, 2016. https://news.usc.edu/107047/experimental-stem-cell-therapy-helps-paralyzed-man-regains-use-of-arms-and-hands/.

13. See, for example, Manley, N. C., Priest, C. A., Denham, J., Wirth, E. D., III, and Lebkowski, J. S., "Human Embryonic Stem Cell–Derived Oligodendrocyte Progenitor Cells: Preclinical Efficacy and Safety in Cervical Spinal Cord Injury," *Stem Cells Translational Medicine* 6, no. 10 (2017): 1917–1929. http://doi.org/10.1002/sctm.17-0065.

14. Keirstead, H. S., Nistor, G., Bernal, G., Totoiu, M., Cloutier, F., Sharp, K., and Steward, O., "Human Embryonic Stem Cell–Derived Oligodendrocyte Progenitor Cell Transplants Remyelinate and Restore Locomotion after Spinal Cord Injury," *Journal of Neuroscience* 25, no. 19 (2005): 4694–4705. https://www.jneurosci.org/content/jneuro/25/19/4694.full.pdf.

15. Priest, C. A., Manley, N. C., Denham, J., Wirth, E. D., III, and Lebkowsi, J. S., "Preclinical Safety of Human Embryonic Stem Cell–Derived Oligodendrocyte Progenitors Supporting Clinical Trials in Spinal Cord

Injury," *Future Medicine* 10, no. 8 (2015): 939–958. https://www.futuremedicine.com/doi/pdfplus/10.2217/rme.15.57.

16. "Collaboration between Lund University Researchers and Novo Nordisk Paves the Way for Large-Scale Cell Therapy against Parkinson's Disease," Lund University, Faculty of Medicine, May 16, 2018. https://www.med.lu.se/english/news_archive/180516_parkinson.

17. "Bayer and Versant Establish iPSC Therapeutics Company BlueRock with $225M," *GEN*, December 12, 2016. http://www.genengnews.com/gen-news-highlights/bayer-and-versant-establish-ipsc-therapeutics-company-bluerock-with-225m/81253536.

18. "Kyoto University Reprograms Stem Cells to Fight Parkinson's in Monkeys: A Breakthrough for Therapy," *Japan Times* (Reuters), August 31, 2007. https://www.japantimes.co.jp/news/2017/08/31/national/science-health/kyoto-university-team-reprograms-stem-cells-fight-parkinsons-disease-monkeys/.

19. Parmar, M.. "Towards stem cell based therapies for Parkinson's disease," *Development* 145, no. 1 (2018): dev156117–4. http://doi.org/10.1242/dev.156117.

20. Williams, S., "Optogenetic Therapies Move Closer to Clinical Use: With a Clinical Trial Underway to Restore Vision Optogenetically, Researchers Also See Promise in Using the Technique to Treat Deafness, Pain, and Other Conditions," *Scientist*, November 16, 2017. https://www.the-scientist.com/?articles.view/articleNo/50980/title/Optogenetic-Therapies-Move-Closer-to-Clinical-Use/.

21. Kashani, A. H., Lebkowski, J. S., Rahhal, F. M., Avery, R. L., Salehi-Had, H., Dang, W., et al., "A Bioengineered Retinal Pigment Epithelial Monolayer for Advanced, Dry Age-Related Macular Degeneration," *Science Translational Medicine* 10, no. 435 (2018): eaao4097. https://stm.sciencemag.org/content/10/435/eaao4097.

22. Mandai, M., Watanabe, A., Kurimoto, Y., Hirami, Y., Morinaga, C., Daimon, T., et al., "Autologous Induced Stem-Cell–Derived Retinal Cells for Macular Degeneration," *New England Journal of Medicine* 376, no. 11

(2017): 1038–1046. http://doi.org/10.1056/NEJMoa1608368 http://doi.org/10.2217/rme-2017-0130.

Chapter 11

1. Merkle, F. T., Ghosh, S., Kamitaki, N., Mitchell, J., Avior, Y., Mello, C., et al. "Human Pluripotent Stem Cells Recurrently Acquire and Expand Dominant Negative P53 Mutations," *Nature* 29, no. 7653 (2017): 1–11. http://doi.org/10.1038/nature22312.

2. For a review of studies on gene sequencing of stem cell cultures, see Martin, U. "Genome Stability of Programmed Stem Cell Products," *Advanced Drug Delivery Reviews* 120 (2017): 108–117. http://doi.org/10.1016/j.addr.2017.09.004.

3. See Mandai et al., "Autologous Induced Stem-Cell-Derived Retinal Cells."

4. See Chakradhar, S. "An Eye to the Future: Researchers Debate Best Path for Stem Cell–Derived Therapies," *Nature Medicine* 22, no. 2 (2016): 116–119. http://doi.org/10.1038/nm0216-116.

5. Takashima, K., Inoue, Y., Tashiro, S., and Muto, K., "Lessons for Reviewing Clinical Trials Using Induced Pluripotent Stem Cells: Examining the Case of a First-in-Human Trial for Age-Related Macular Degeneration," *Regenerative Medicine* 13, no. 2 (2018): 123–128. http://doi.org/10.2217/rme-2017-0130.

6. Note there are two different nomenclatures here: sometimes major histocompatibility complex (MHC) antigens are referred to as "human leukocyte antigens" (HLAs), but both are referring to the same thing.

7. Opelz, G., and Döhler, B., "Pediatric Kidney Transplantation: Analysis of Donor Age, HLA Match, and Posttransplant Non-Hodgkin Lymphoma: A Collaborative Transplant Study Report," *Transplantation* 90, no. 3 (2010): 292–297. http://doi.org/10.1097/TP.0b013e3181e46a22.

8. Gourraud, P.-A., Gilson, L., Gilson, Girard, M., and Peschanski, M. "The Role of Human Leukocyte Antigen Matching in the Development

of Multiethnic 'Haplobank' of Induced Pluripotent Stem Cell Lines," *Stem Cells* 30, no. 2 (2012): 180–186. http://doi.org/10.1002/stem.772.

9. See Center for iPS Cell Research and Application (CiRA), Kyoto University, website: http://www.cira.kyoto-u.ac.jp/e/index.html.

10. See Global Alliance for induced Pluripotent Stem Cell Therapies (GAiT), Edinburgh, website: http://www.gait.global/about-gait/.

11. Sullivan, T., "A Tough Road: Cost to Develop One New Drug Is $2.6 Billion; Approval Rate for Drugs Entering Clinical Development Is Less Than 12%," *Policy & Medicine*, last updated March 21, 2019. https://www.policymed.com/2014/12/a-tough-road-cost-to-develop-one-new-drug-is-26-billion-approval-rate-for-drugs-entering-clinical-de.html.

12. Bravery, C. A., "Do Human Leukocyte Antigen–Typed Cellular Therapeutics Based on Induced Pluripotent Stem Cells Make Commercial Sense?," *Stem Cells and Development* 24, no. 1 (2015): 1–10. http://doi.org/10.1089/scd.2014.0136.

13. Gornalusse, G. G., Hirata, R. K., Funk, S. E., Riolobos, L., Lopes, V. S., Manske, G., et al., "HLA-E-Expressing Pluripotent Stem Cells Escape Allogeneic Responses and Lysis by NK Cells," *Nature Biotechnology* 35, no. 8 (2017): 765–772. http://doi.org/10.1038/nbt.3860.

14. Park, K. I., Teng, Y. D., and Snyder, E. Y., "The Injured Brain Interacts Reciprocally with Neural Stem Cells Supported by Scaffolds to Reconstitute Lost Tissue," *Nature Biotechnology* 20, no. 11 (2002): 1111–1117. http://doi.org/10.1038/nbt751.

15. Yang, L., Chueng, S.-T. D., Li, Y., Patel, M., Rathnam, C., Dey, G., et al., "A Biodegradable Hybrid Inorganic Nanoscaffold for Advanced Stem Cell Therapy," *Nature Communications* 9, no. 1; article no. 3147: 1–14. http://doi.org/10.1038/s41467-018-05599-2.

16. For a more detailed discussion of these issues, see Stevens, K. R., and Murry, C. E., "Human Pluripotent Stem Cell–Derived Engineered Tissues: Clinical Considerations," *Stem Cell* 22, no. 3 (2018): 294–297. http://doi.org/10.1016/j.stem.2018.01.015.

Chapter 12

1. Davis, R. L., Weintraub, H., and Lassar, A. B., "Expression of a Single Transfected cDNA Converts Fibroblasts to Myoblasts," *Cell* 51, no. 6 (1987) 987–1000. https://doi.org/10.1016/0092-8674(87)90585-X.

2. Jopling, C., Boué, S., and Belmonte, J. C. I., "Dedifferentiation, Transdifferentiation and Reprogramming: Three Routes to Regeneration, *Nature Reviews Molecular Biology* 12, no. 2 (2011): 79–89. http://doi.org/10.1038/nrm3043.

3. Vierbuchen, T., Ostermeier, A., Pang, Z. P., Kokubu, Y., Sudhof, T. C., and Wernig, M., "Direct Conversion of Fibroblasts to Functional Neurons by Defined Factors," *Nature* 463, no. 7284 (2010): 1035–1041. http://doi.org/10.1038/nature08797.

4. Buffo, A., Vosko, M. R., Erturk, D., Hamann, G. F., Jucker, M., Rowitch, D. H., and Gotz, M., "Expression Pattern of the Transcription Factor Olig2 in Response to Brain Injuries: Implications for Neuronal Repair," *Proceedings of the National Academy of Sciences* 102, no. 50 (2005): 18183–18188. https://www.pnas.org/content/102/50/18183; Heins, N., Malatesta, P., Cecconi, F., Nakafuku, M., Tucker, K. L., Hack, M. A., et al., "Glial Cells Generate Neurons: The Role of the Transcription Factor Pax6," *Nature Neuroscience* 5, no. 4 (2002): 308–315. http://doi.org/10.1038/nn828.

5. di Val Cervo, P. R., Romanov, R. A., Spigolon, G., Masini, D., Martín-Montañez, E., Toledo, E. M., et al., "Induction of Functional Dopamine Neurons from Human Astrocytes *in Vitro* and Mouse Astrocytes in a Parkinson's Disease Model," *Nature Biotechnology* 35, no. 5 (2017): 444–452. http://doi.org/10.1038/nbt.3835.

6. Victor, M. B., Richner, M., Hermanstyne, T. O., Ransdell, J. L., Sobieski, C., Deng, P.-Y., et al., "Generation of Human Striatal Neurons by MicroRNA-Dependent Direct Conversion of Fibroblasts," *Neuron* 84, no. 2 (2014): 311–323. http://doi.org/10.1016/j.neuron.2014.10.016.

7. Francis Crick, as quoted in Hall, S. S., "Hidden Treasures in Junk DNA: What Was Once Known as Junk DNA Turns Out to Hold

Hidden Treasures, Says Computational Biologist Ewan Birney," *Scientific American*, October 1, 2012. https://www.scientificamerican.com/article/hidden-treasures-in-junk-dna/.

8. See, for example, Lee, J.-H., Mitchell, R. R., McNicol, J. D., Shapovalova, Z., Laronde, S., Tanasijevic, B., et al., "Single Transcription Factor Conversion of Human Blood Fate to NPCs with CNS and PNS Developmental Capacity," *Cell Reports* 11, no. 9 (2015): 1367–1376. http://doi.org/10.1016/j.celrep.2015.04.056.

9. See, for example, Gascón, S., Masserdotti, G., Russo, G. L., and Götz, M., "Direct Neuronal Reprogramming: Achievements, Hurdles, and New Roads to Success," *Stem Cell* 21, no. 1 (2017): 18–34. http://doi.org/10.1016/j.stem.2017.06.011.

10. Heinrich, C., Bergami, M., Gascón, S., Lepier, A., Viganò, F., Dimou, L., et al., "Sox2-Mediated Conversion of NG2 Glia into Induced Neurons in the Injured Adult Cerebral Cortex," *Stem Cell Reports* 3, no. 6 (2014): 1000–1014. https://www.sciencedirect.com/science/article/pii/S2213671114003294.

Index

Adrenal medulla transplants, 83–84, 92, 94, 105
Altman, Joseph, 50–57, 63
Alvarez-Buylla, Arturo, 56
Alzheimer's disease, 2–4, 32, 65
Amyotrophic lateral sclerosis (ALS), 21, 188–190
Anderson, Aileen, 126
Apoptosis, 24
Asterias, 128, 198–200
Astrocytes, 22, 25, 28, 69, 193
Autologous therapies, 38, 153–156, 162, 209, 215–217
Avian brain, 64, 7

Basal ganglia, 30, 70
Batten's disease, 115–116, 120
Biotechnology, 1, 5–6, 8–9, 189
Björklund, Anders, 91–92, 98
Blastema, 232
Bliss, Tim, 58
Bone marrow, 7, 33–35, 37–42, 47, 57, 80, 111, 137, 139, 150, 154–155, 157, 167, 230
Brain networks, 26–29

California Institute of Regenerative Medicine (CIRM), 120, 127, 185, 198
Cancer, 2, 19, 37–38, 62, 130, 131, 149, 211–212, 213–215, 225, 228
Cardiomyocytes, 49, 232
Cell cycle, 114, 130–131, 213
Cerebellum, 28, 29, 30
Cerebral cortex, 26–28, 58, 70, 71, 75, 92, 107, 121, 131, 193–194
Chang, Hannah, 43–44
Chirba, Mary Ann, 157–158, 164
Clinical trials, 8, 85–86, 88, 91, 105, 120–122, 126, 127, 190
  failure to publish, 126–127
  failure rate of, 3–4
  motor neuron disease (*see* Motor neuron disease)
  Parkinson's disease (*see* Parkinson's disease: clinical trials)
  patient consent for, 89, 126, 160–161, 198

Clinical trials (cont.)
  spinal cord injury (*see* Spinal cord injury)
  stroke (*see* Stroke: clinical trials)
ClinicalTrials.gov, 127, 148
Coffey, Pete, 206–209, 239
Combination products, 227
Concussion, 19
Corpus striatum, 106–108, 139–141
Corticospinal tract, 118
Cowan, W. Maxwell, 58
Cummings, Brian, 119–120, 126, 204
Cytokines, 23
  EGF, 114
  erthyropoietin, 41–43
  FGF, 114, 121 174
  GM-CSF, 41–43
  LIF, 174–175
  TGFβ, 174

da Cruz, Lyndon, 206
Depression, 66–67, 87
Diaschisis, 32
Diphtheria toxin killing, 120
Direct Reprogramming, 229–237
Disease modeling, 187–190
Doetsch, Fiona, 78
Dolly the Sheep, 172, 178–179
Dunnett, Steve, 91

Embryonic stem cells (ES cells), 170–171, 174–177, 183–185, 187, 191, 194, 196–199, 202, 207, 209, 217
Ethics, 176

Equipoise, 159–161
Eriksson, Peter, 63
European Medicines Agency, 151, 177
Excitotoxicity, 22, 30, 188

Fleck, Ludwig, 57
Food and Drug Adminstration (FDA), 5, 85, 86, 102, 151, 155–158, 164, 177, 196–198
Freed, Curt, 91, 96, 100
Frisén, Jonas, 63

Garfield, Stephanie, 157–158, 164
Genome editing. *See* Induced pluripotent stem cells (iPS cells): genome editing
Geron Corporation, 8, 196–198
Glia, 2, 28, 31, 34, 49, 100, 110, 112, 131, 136–138, 167, 193
  relationship to NSCs, 69–70, 236–237
Glial limiting membrane, 25
Goldstein, Larry, 126–127
Gould, Elizabeth, 67
Grey, Jeffrey, vii, 58
Growth factors. *See* Cytokines
Gurdon, John, 78, 179, 183

Hawking, Stephen, 162
Heart disease, xii, 2
Hematopoiesis, 33–37
Hematopoietic stem cells (HSC), 33, 34–35, 37–43, 71, 111, 113, 139

Hepatocytes, 49
Hippocampus, 26, 51–52, 57–59, 62–64, 67, 68, 70
  dentate gyrus, 34, 51–53, 57–66, 68–69, 87, 111
  pattern separation, 60
  perforant pathway, 57–58
Hodges, Helen, xiii
Hospital exemption, 153–154, 157
Human Fertilization and Embryology Act, 176
Huntingtin, 106–107
Huntington's disease, 11, 65, 70, 105–108, 109, 196
Hwang Woo-Suk, 172

Immunomodulation, 141, 150
Induced pluripotent stem cells (iPS cells), 35, 183–186
  clinical applications, 209–211, 215–217
  comparability, 221–224
  epigenetics, 184–5
  genome editing, 225–226
  haplobanks, 220–226
Inner cell mass, 34, 168–170, 231
International Society for Stem Cell Research (ISSCR), 127, 161, 164
  patient handbook, 161
In vitro fertilization (IVF), 6–7, 15, 173–175
In vivo reprogramming, 236
Isacson, Ole, 98

Kaplan, Michael, 53–57
Kashani, Amir, 207–209
Keller, Gordon, 203
Kirkeby, Agnete, 203
Kuhn, Ludwig, 57

Laskey, Ron, 178
Laurie, Graeme, 177
Lindvall, Ole, 73–77, 91–92, 137
Lømo, Terje, 58
London Project to Cure Blindness, 127, 206
Long Term Potentiation (LTP), 58–59
Lou Gehrig's disease. *See* Amyotrophic lateral sclerosis (ALS)
Lower vertebrates, 15, 49, 68

Macklis, Jeff, 74–79, 119, 137
Major Histocompatibility Complex (MHC), 218–219, 225
McCormack, Kevin, 120
McCulloch, Ernest, 35
McKewen, Bruce, 67
Memory, 58–60, 64, 66, 70–71
Mesenchymal stromal cells (MSCs), 34, 137–138, 150–151
MHRA, xiii, 86, 134
Microglia, 23–24, 28, 141
Mode of action, 34, 67, 85–88, 142, 149–152, 202–203
Mood, 58, 66–67, 107
Monuki, Edwin, 120

Morris, Richard, 59
Morris Water Maze, 59
Motor neuron disease (MND). *See* Amyotrophic lateral sclerosis (ALS)
Multiple sclerosis, 38, 125, 162

Necrosis, 24
Neural networks, 26–30
Neural stem cells (NSCs), 34, 35, 39, 45, 47, 52, 53, 58, 69–71, 79–80, 109, 111, 193
  conditional immortalisation of, 130–132
  neurospheres, 114–115
Neurectoderm, 110, 186
Neurodegeneration, 32, 107–108, 115–116, 141
Neurogenesis, 48, 51, 55, 55–61, 61–67, 68–78
  death of a dogma, 47–51
  in songbirds, 54–55
  stress, effect of, 67
  subependymal zone (SEZ), 68–71, 74–76, 111
Neurons
  dopaminergic, 2, 80–89, 92–93, 96–100, 102–13, 105–107, 111, 202–205, 233
  interneuron, 28, 29, 71, 74
  markers for, 56–57, 62–63, 74, 77, 200, 203
  medium spiny, 74, 92, 106–108, 235
  motor, 118–119, 121, 186, 188–189, 192
  projection, 27–28, 55, 71, 74–75
  pyramidal, 26–29, 75–76, 107, 118, 121
Noise, 43–44, 187
Notch signaling pathway, 132, 138
Nottebohm, Fernando, 54–56
Nowakowski, Richard, 49

Olfactory bulb, 26, 58, 70–71, 76
Oncogenes, 130
Optogenetics, 78, 204–205

Parkinson's disease, xii, 2, 20, 65, 70, 80–89, 91–105, 106–108, 111, 142, 161, 192, 200, 202–205
  alpha-synuclein, 102–103
  animal model, 88–92, 204
  cell therapy, 84–89, 161, 202–205
  clinical trials, 91–101
  deep brain stimulation, 101–102
  fetal transplants, 104–105
  Lewy bodies, 102–103
  pathology, 102–104
  TRANSEURO, 108
Parmar, Malin, 202
Peschanski, Marc, 220
PISCES clinical trial, xiii, 134
Placebo, 88–89, 91, 93, 97, 99, 127, 135, 142–144, 159–161
Pluripotent cells, tumorogenicity of, 214–215
Progenitor cells, 36–37
  dopaminergic, 203

hematopoietic, 34, 39, 40–41, 43
neural, 48, 57, 60, 69, 100,
  110–111, 111–114, 129–132,
  186, 193, 200, 235
oligodendrocyte (OPCs), 197,
  199, 221
proof of concept, 98–101, 202
radial glia, 111

Ramón y Cajal, Santiago, 47, 73
Reeve, Christopher, 117
Retina, 105, 122–125, 196,
  205–211, 216
  age-related macular
    degeneration (ARMD), 116,
    123–124, 128, 147, 205, 208
  Bruch's membrane, 123, 207
  clinical trials, 206, 215
  pigment epithelium, 122–125,
    206–27, 209, 211, 221, 223
  retinitis pigmentosa, 123
  Stargardt's disease, 208
Reynolds, Brent, 114
Rostral migratory stream, 70–71

Scharff, Constance, 78–79
Schwartz, Steven, 207–208
Serotonin, 87, 100
Severe combined
  immunodeficiency (SCID),
  113
Sexual dimorphism, 55
Sinden, John, xii, 131, 239
Spheramine, 115
Spinal cord injury, 116, 120, 124,
  128, 148, 164, 196–197, 199

Stem cell
  competence, 42–44
  niche, 37, 40–41, 57, 68–69, 74,
    76, 78–80, 139, 154
  properties, 32–33, 38–39, 69,
    111–113, 132
Stem cell tourism, 16, 148, 161
Stroke, 9, 22, 27, 77, 147, 205,
  206, 210
  advertising campaign for, 20
  animal model, 4–5, 13, 133–134
  clinical trials, 132, 133, 138,
    142–144
  hemorrhagic, 143
  ischemic, 2, 20, 23, 134
  pathology, 11–12, 73–74, 116
  penumbra, 24–25, 30
Studer, Lorenz, 134, 203–204,
  226
Substantia nigra, 81–82, 85, 92
Swift, Teresa, 16

Takahashi, Jun, 203
Takahashi, Kazutoshi, 35, 180,
  181–184
Testosterone, 54–55
Thomson, James, 183, 197, 229
Thrombolytic agents, 23
Till, James, 35
Tissue rejection, 95, 100. 216–217
Transcription factor, 42–43,
  183–184, 230, 233–234, 237
Transdifferentiation, 133, 137–138,
  232
Traumatic brain injury, 11, 125,
  143

Tritiated thymidine, 50–51, 53, 55–56, 63
Tumorogenicity, 214–215

UK Stem Cell Bank, 176

Ventral mesencephalon, 92

Warnock Report, 176
Weijer, Charles, 161
Weiss, Paul, 53
Weiss, Sam, 114
West, Michael, 198

Yamanaka, Shinja, 17, 35, 167, 180–184, 229, 232

Zhu, Jianhong, 114